ORIGINAL PAPERS

BY

JOHN HOPKINSON

ORIGINAL PAPERS

BY THE LATE

JOHN HOPKINSON, D.Sc., F.R.S.

VOL. II.

SCIENTIFIC PAPERS

EDITED BY

B. HOPKINSON, B.Sc.

CAMBRIDGE:

AT THE UNIVERSITY PRESS

1901

CAMBRIDGE
UNIVERSITY PRESS

University Printing House, Cambridge CB2 8BS, United Kingdom

Cambridge University Press is part of the University of Cambridge.

It furthers the University's mission by disseminating knowledge in the pursuit of education, learning and research at the highest international levels of excellence.

www.cambridge.org
Information on this title: www.cambridge.org/9781107455993

© Cambridge University Press 1901

First published 1901
First paperback edition 2014

A catalogue record for this publication is available from the British Library

ISBN 978-1-107-45599-3 Paperback

CONTENTS OF VOL. II.

18.

THE RESIDUAL CHARGE OF THE LEYDEN JAR.

[From the *Philosophical Transactions of the Royal Society*,
Vol. CLXVI. Part II. pp. 489—494, 1876.]

Received February 24,—*Read March* 30, 1876.

1.   SUPPOSE that the state of a dielectric under electric force*
is somewhat analogous to that of a magnet, that each small portion
of its substance is in an electropolar state. Whatever be the
ultimate physical nature of this polarity, whether it arises from
conduction, the dielectric being supposed heterogeneous (see
Maxwell's *Electricity and Magnetism*, vol. I. arts. 328—330), or
from a permanent polarity of the molecules analogous to that
assumed in Weber's theory of induced magnetism, the potential
at points external to the substance due to this electropolar state
will be exactly the same as that due to a surface distribution of
electricity, and its effect at all external points may be masked by
a contrary surface distribution. Assume, further, that dielectrics
have a property analogous to coercive force in magnetism, that
the polar state does not instantly attain its full value under
electric force, but requires time for development, and also for
complete disappearance, when the force ceases. The residual
charge may be explained by that part of the polarization into

---

* To define the electric force within the dielectric it is necessary to suppose a
small hollow space excavated about the point considered; the force will depend
on the form of this space; but it is not necessary for the present purpose to decide
what form it is most appropriate to assume.

H. II.   1

which time sensibly enters. A condenser is charged for a time, the dielectric gradually becomes polarized; on discharge the two surfaces of the condenser can only take the same potential if a portion of the charge remain sufficient to cancel the potential, at each surface, of the polarization of the dielectric. The condenser is insulated, the force through the dielectric is insufficient to permanently sustain the polarization, which therefore slowly decays; the potentials of the polarized dielectric and of the surface charge of electricity are no longer equal, the difference is the measurable potential of the residual or return charge at the time. It is only necessary to assume a relation between the electric force, the polarization measured by the equivalent surface distribution, and the time. For small charges a possible law may be the following:—For any intensity of force there is a value of the polarization, proportional to the force, to which the actual polarization approaches at a rate proportional to its difference therefrom. Or we might simply assume that the difference of potential E of the two surfaces and the polarization are connected with the time by two linear differential equations of the first order. If this be so, E can be expressed in terms of the time t during insulation by the formula $E = (A + Be^{-\mu t})\, \epsilon^{-\lambda t}$, where λ and μ are constants for the material, and A and B are constants dependent on the state of the dielectric previous to insulation. It should be remarked that λ does not depend alone on the conductivity and specific inductive capacity, as ordinarily deter- mined, of the material, but also on the constants connecting polarization with electric force. Indeed if the above view really represent the facts, the conductivity of a dielectric determined from the steady flow of electricity through it measured by the galvanometer will differ from that determined by the rate of loss of charge of the condenser when insulated.

2. A Florence flask nearly 4 inches in diameter was carefully cleansed, filled with strong sulphuric acid, and immersed in water to the shoulder. Platinum wires were dipped in the two fluids, and were also connected with the two principal electrodes of the quadrant electrometer. The jar was slightly charged and insulated, and the potentials read off from time to time. It was found (1) that even after twenty-four hours the percentage of loss per hour continued to decrease, (2) that the potential could not be expressed as a function of the time by two exponential terms. But the

latter fact was more clearly shown by the rate of development of the residual charge after different periods of discharge, which put it beyond doubt that if the potential is properly expressed by a series of exponential terms at all, several such terms will be required.

The following roughly illustrates how such terms could arise. Glass may be regarded as a mixture of a variety of different silicates; each of these may behave differently under electric force, some rapidly approaching the limiting polarity corresponding to the force, others more slowly. If these polarities be assumed to be n in number, they and E may be connected with the time by $n + 1$ linear differential equations. Hence during insulation E would be expressed in the form $\Sigma_0{}^n A_r \epsilon^{-\lambda_r t}$. Suppose now a condenser be charged positively for a long time, the polarization of all the substances will be fully developed; let the charge be next negative for a shorter time, the rapidly changing polarities will change their sign, but the time is insufficient to reverse those which are more sluggish. Let the condenser be then discharged and insulated, the rapid polarizations will decay, first liberating a negative charge; but after a time the effect of the slow terms will make itself felt and the residual charge becomes positive, rises to a maximum, and then decays by conduction. This inference from these hypotheses and the form of the curve connecting E with t for a simple case of return charge is verified in the following experiments.

3. A flask was immersed* in and filled with acid to the shoulder. Platinum electrodes communicated with the electrometer as before. The flask was strongly charged positive at 5.30 and kept charged till 6.30, then discharged till 7.8 and negatively charged till 7.15, when it was discharged and insulated. The potential was read off at intervals till 8.20. The abscissæ of curve A (Fig., p. 4) represent the time from insulation, the ordinates the corresponding potentials, positive potentials being measured upwards. It will be seen that a considerable negative charge first appeared, attaining a maximum in about five minutes; it then decreased, and the potential was *nil* in half an hour; the main positive return charge then came out, and was still rapidly increasing at 8.20, when the

---

* Acid on both sides of the dielectric, that there might be no electromotive force from the action of acid on water either through or over the surface of the glass.

flask was again discharged. At 8.39 the same flask was charged negatively till 8.44, then discharged and charged positively for

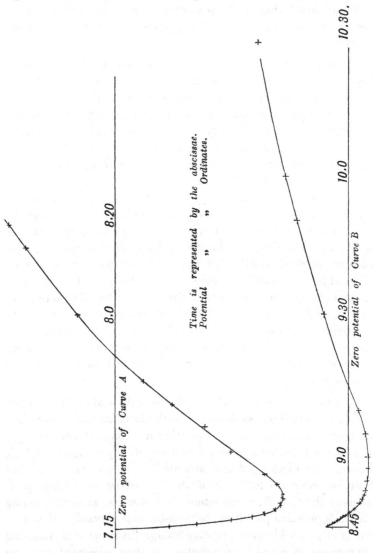

45 seconds, insulated 15 seconds and discharged, and finally insulated at 8.45. Curve *B* represents the subsequent potentials. It is seen that the return charge twice changes sign before it assumes its final character. The experiment was several times repeated with similar results.

Sir William Thomson has informed the author, since these experiments were tried, that he himself performed similar experiments many years ago, and showed them as lecture illustrations in his Class in the University of Glasgow, but never otherwise published them, proving that the charges come out of the glass in the inverse order to that in which they go in†.

4.   When steel is placed in a magnetic field, mechanical agitation accelerates the rapidity with which its magnetic polarity is developed.   Again, vibration reduces the magnetism of a magnet, or, so to speak, shakes its magnetism out.   This would suggest, on the present hypothesis, that vibration would accelerate changes in the electric polarity of a dielectric, or shake down polarization and liberate residual charge.   The following experiments verify this anticipation.   The arrangement was as in (3). The flask was strongly charged for some hours, discharged at 4.45 P.M., and kept with the two coatings connected by a platinum wire, except for a few moments at a time, to ascertain the rate at which the polarization was decaying, till 9.48, when the flask was insulated and the number of seconds observed in which the potential rose to 100, 200, 300, and 500 divisions of the scale of the quadrant electrometer, every thing being as steady as possible.   The flask was then discharged, again insulated at 10.18, and the development of the charge observed, the neck of the flask being sharply tapped during the whole time.   The experiment was repeated quiet at 10.48, with tapping at 11.16.   Column I. gives the time of beginning the observation, II., III., IV., and V. the number of seconds in which charges 100, 200, 300, 500

I.	II.	III.	IV.	V.
9.48	118	240	367	624
10.18	80*	140*	185*	320*
10.48	140	285	440	750
11.16	120*	210*	310*	540*

† These results are closely analogous to those obtained by Boltzmann for torsion (*Sitzungsberichte der k. Akad. der Wiss. zu Wien*, Bd. LXX. Sitzung 8, Oct. 1874).   From his formulæ it follows that if a fibre of glass is twisted for a long time in one direction, for a shorter time in the opposite direction, and is then released, the set of the fibre will for a time follow the last twist, will decrease, and finally take the sign of the first twist.

developed respectively. The periods of tapping are marked with an asterisk.

The effect may appear small; but it must be remembered that, the flask containing and being immersed in sulphuric acid to the shoulder, the vibration caused by tapping the neck could be but small, and could scarcely penetrate to the lower part of the flask. The experiment was subsequently repeated with the same flask and with a similar result; but it was further found that the effect of tapping was more marked when the periods during which the flask was strongly charged and discharged were long than when they were short. For example, when the flask was charged half an hour, then discharged five minutes, the effect of tapping was very slight although unmistakable. That portion of the return charge which comes out slowly is more accelerated by vibration than that which comes out fast. A comparison of the rates at 10.18 and 11.16 of the above table also shows that a flask which has been tapped is less susceptible to the effect of tapping than it was before it was touched. In some cases also it was noticed that if three observations were made, the first quiet, the second tapped, and the third quiet, the third charge came out more rapidly than the first. The last experiment on tapping below illustrates both of these points.

A flask was mounted as before, strongly charged at 12 o'clock, discharged at 3, and remained discharged till 5.15, when it was insulated, and the time which the image took to traverse 200 divisions was noted; after passing that point the flask was again discharged. The first column gives the instant of insulation, the second the time of covering 200 divisions. The observations without mark were made with the flask untouched, in those marked * it was sharply tapped all over with a glass rod dipping in the acid, whilst in those marked † the rod was muffled with a piece of india-rubber tubing.

Time of insulation.		Time occupied in traversing 200 divisions of the scale.	
h.	m.	min.	secs.
5	15	1	23
5	18	1	23
*5	21		40
5	24	1	17
†5	27		48
5	30	1	27

	Time of insulation. h. m.	Time occupied in traversing 200 divisions of the scale. min. secs.
Remained discharged till	5 $46\frac{1}{2}$	1 25
	†5 $49\frac{1}{2}$	54
	5 $51\frac{1}{2}$	1 24
	*5 54	1 3
	5 56	1 24
Remained discharged till	6 39	2 $7\frac{1}{2}$
	*6 43	1 54
	6 47	2 8
	†6 51	2 2
	6 55	2 13
Remained discharged till	8 51	2 11
	†8 55	57
	8 58	2 12
	*9 2	1 2
	9 5	2 14
	†9 9	54
	9 12	2 17

The same flask was strongly charged at 9.15 in the evening and discharged at 9 the following morning, and remained so till 7.13 in the evening, when the following observations of the time of traversing 100 divisions were made :—

	Time of insulation. h. m.	Time occupied in traversing 100 divisions of the scale. min. secs.
	7 13	2 43
	†7 18	1 35
	7 21	2 35
	*7 25	1 53
	7 28	2 27
	†7 32	1 46
	7 35	2 26
	*7 39	1 49
	7 43	2 25

The result here was less than the author expected, considering the long period of discharge and the considerable effect obtained in the previous experiment; this may perhaps be due to change of temperature, or perhaps to a difference in the vigour with which the flask was tapped*.

5. When a charge is given to an insulated flask, owing to polarization the percentage of loss per minute continuously diminishes towards a limiting value. When the flask is charged, discharged, and insulated, one would expect that after attaining a maximum potential the rate of loss would steadily *increase* towards the same limiting value as in the former case. The following experiment shows that this is not always the case.

A flask of window-glass, much more conductive than the Florence flask, was mounted as in (3) and (4). It was charged, and the charge maintained for three quarters of an hour, then discharged for a quarter of an hour, and insulated. In four minutes the charge attained a maximum value 740. In fifteen minutes the potential was 425, in twenty minutes 316, giving a loss in five minutes of 26 per cent. In thirty minutes it was 186, and in thirty-five minutes 146, a loss of 21½ per cent. The intermediate and subsequent readings of the same series showed a steady decrease to as little as 15 per cent. The experiment was repeated with the same flask, but with shorter periods of discharge and with a similar result.

6. Although the above view is only proposed as a provisional working hypothesis, some suggestions which it indicates may be pointed out.

Temperature has three effects on the magnetic state of iron or steel:—(1) changes of temperature cause temporary changes in the intensity of a magnet; (2) temperature affects the "permeability" of a magnet; at a red heat iron is no longer sensibly magnetic; (3) a rise of temperature reduces coercive force.

It may be expected that the polarity of dielectrics may also be affected in three analogous ways:—(1) a sudden change of temperature might directly and suddenly affect the polarity (an example of this we have in the phenomena of pyro-electricity);

---

* It is recorded by Dr Young that an electrical jar may be discharged either by heating it or by causing it to sound by the friction of the finger

(2) the constant expressing the ratio of limiting polarity to electromotive force may depend on temperature; and (3) temperature may alter the constant, expressing the rate at which polarity approaches its limiting value for a given force, as it is known to alter the specific conductivity. Mr Perry's experiments show that temperature does affect the polarization of dielectrics, but in which way does not appear.

Sir William Thomson (papers on Electrostatics and Magnetism, art. 43) explains specific inductive capacity by a polarization of the dielectric following the same formal laws as magnetism. It is only necessary to introduce time into that explanation as here proposed to enable it to cover the phenomena of residual charge. Again (see Nichol's *Cyclopædia*), Sir William Thomson explains the phenomena of pyro-electricity by supposing that every part of the crystal of tourmaline is electropolar, that temperature changes the intensity of its polarity, and that this polarity is masked by a surface distribution of electricity supplied by conduction over the surface or otherwise. We have, then, in tourmaline an analogue to a rigidly magnetized body, in glass or other dielectrics analogues to iron having more or less coercive force.

19.

RESIDUAL CHARGE OF THE LEYDEN JAR.— DIELECTRIC PROPERTIES OF DIFFERENT GLASSES.

[From the *Philosophical Transactions of the Royal Society*, Vol. CLXVII. Part II. pp. 599—626, 1877.]

Received November 30, 1876,—*Read January* 18, 1877.

I. BEFORE proceeding to comparative experiments on different glasses, it appeared desirable to verify experimentally the two following propositions:—

(*a*) If two jars be made of the same glass but of different thicknesses, if they be charged to the same potential for equal times, discharged for equal times and then insulated, the residual charge will after equal times have the same potential in each. In experiments in which potentials and not quantities of electricity are measured the thickness of the jar may be chosen arbitrarily, nor need any inconvenience be feared from irregularities of thickness.

(*b*) Residual charge is proportional to exciting charge.

These propositions may be included in one law—that superposition of simultaneous forces is applicable to the phenomena of residual charge.

To verify (*a*) two flasks were prepared of the glass afterwards referred to as No. 1. One was estimated to be about 1 millim., the other 6 or 7 millims. thick. These were cleansed and insulated in the usual way by filling with strong sulphuric acid without soiling the neck of the flask. They were placed in the same basin of water, which was electrically connected with the outside of the quadrant electrometer. The interiors of the flasks were respectively connected with the two quadrants; they were also connected together by a wire which could at any instant be removed. One Daniell's element gave a deflection of 69 scale-divisions. The two flasks were charged together with 48 elements for some minutes, and it was observed that the equal charge of the two quadrants did not deflect the needle. The flasks were discharged for 15 or 20 seconds and insulated, still connected. The connecting wire was then removed, and the subsequent movement of the image observed. If left undisturbed a maximum of about 20 divisions of the scale was attained. But usually the deflection in from 20 to 30 seconds reaching 10 divisions, the thick flask was discharged, and the image was driven from the scale, showing that at that time the potential of either flask was represented by more than 500 scale-divisions, and hence that the difference between them was less than 2 per cent. of either of them. When the charge was negative the error was in favour of the thin flask. This is in complete accord with anomalous results subsequently obtained with the same glass. Correcting for this peculiarity of the glass we may conclude that the law is verified within the limits of these experiments.

The second proposition was confirmed with two different glasses; but the results in one case are not quite accordant, possibly owing to variations of temperature, or to slight unremoved effects of previous chargings; but the irregularities indicate no continuous deviation from the law. In these and all the subsequent experiments the flasks were blown as thin as possible in the body, but with thick necks, the neck being thick that the capacity of any zone might be small.

Flask of optical soft crown, No. 5. The electrometer reads 28½ for one Daniell's element. The charging in each case lasted some hours, the discharge 30 seconds. The flask was then insulated and remained insulated; the residual charge was read off from time to time. Column I. gives the time in minutes from insulation; II., III., IV., V., the readings at those times, the exciting electromotive force being respectively that of 48, 48, 24, and 12 elements of the battery.

I.	II.	III.	IV.	V.
1	90	92	46	....
5	218	225	103	51
15	....	344	160	79
30	....	423	197	99
60	478	462	226	114
120	492	....	233	120

Flask of blue glass, No. 2.   The reading of the electrometer for one element was 69 divisions.   The charge in each case lasted 10 minutes, the discharge 30 seconds; the flask was then insulated.   Column I. gives the time from insulation in minutes; II., III., IV., V., the potentials at those times when the batteries which had been employed were respectively 48, 12, 3, and 1 Daniell's elements.

I.	II.	III.	IV.	V.
$\frac{1}{2}$	414	102	$26\frac{1}{2}$	9
Maximum potential	472	$117\frac{1}{2}$	$30\frac{1}{4}$	10
$1\frac{1}{2}$	456	114	$29\frac{1}{2}$	10
$2\frac{1}{2}$	385	96	$24\frac{3}{4}$	$8\frac{1}{4}$
$4\frac{1}{2}$	256	65	$16\frac{1}{2}$	6
$9\frac{1}{2}$	120	....	$8\frac{1}{2}$	3

The agreement in this case, all the experiments being made on the same day, is fairly satisfactory.

II. The following method of treating the question of residual charge was suggested to the author by Professor Clerk Maxwell; it is essentially similar to that used by Boltzmann for the after-effects of mechanical strain (" Zur Theorie der elastischen Nach-wirkung," aus dem lxx. Bande der *Sitz. der k. Akad. der Wissensch. zu Wien*, II. Abth. Oct. Heft, Jahrg. 1874).

Let L be the couple tending to twist a wire or fibre about its axis, θ_t the whole angle of torsion at time t; then L at time t depends upon θ_t, but not wholly on θ_t, for the torsion to which the wire has been submitted at all times previous to t will slightly affect the value of L. Assume only that the effects of the torsion at all previous times can be superposed. The effect of a torsion $\theta_{t-\omega}$ at a time ω before the time considered, acting for a short time $d\omega$, will continually diminish as ω increases; it may be expressed by $-\theta_{t-\omega} f(\omega) d\omega$, where $f(\omega)$ is a function of ω, which diminishes as ω increases. Adding all the effects of the torsion at all times, we have

$$L = a\theta_t - \int_0^\infty \theta_{t-\omega} f(\omega)\, d\omega.$$

In the case of a glass fibre Boltzmann finds that $f(\omega) = \dfrac{A}{\omega}$, where A is constant for moderate value of ω, but decreases when ω is very great.

The after-effects of electromotive force on a dielectric are very similar; to strain corresponds electric displacement, to stress electromotive force. Let x_t be the potential at time t as measured by the electrometer, and y_t the surface-integral of electric displace-ment divided by the instantaneous capacity of the jar; then, assuming only the law of superposition already proved to be true for simultaneous forces, we may write

$$x_t = y_t - \int_0^\infty y_{t-\omega}\, \phi(\omega)\, d\omega, \quad\ldots\ldots\ldots\ldots\ldots (1)$$

where $\phi(\omega)$ is a function decreasing as ω increases. This formula is precisely analogous to that of Boltzmann; but in the case of a glass jar the capacity of which is too small to give continuous currents, it is not easy to measure y_t; hence it is necessary to make x_t the independent variable. From the linearity of the equation (1) as regards $x_t y_t$ and the value of $y_{t-\omega}$ for each value of ω, and from the linearity of the equation expressing $x_{t-\omega}$ for each value of ω, it follows that

$$y_t = x_t + \int_0^\infty x_{t-\omega} \, \psi(\omega) \, d\omega, \quad\quad\quad\quad (2)$$

where $\psi(\omega)$ decreases as ω increases.

The statement of equations (1) and (2) could be expressed in the language of action at a distance and electrical polarization of the glass, y_t being replaced by the polarization as measured by the potential of the charge which would be liberated if the polarization were suddenly reduced to zero, the jar being insulated. It should be noted that the view of this subject adopted by the author in the previous paper* can be included in equation (2) by assuming that $\psi(\omega)$ is the sum of a series of exponentials.

If $\psi(\omega)$ is determined for all values of ω, the properties of the glass, as regards conduction and residual charge, are completely expressed.

Suppose that in equation (2) $x_t = 0$ till $t = 0$, and that after that time $x_t = X$ a constant,

$$y_t = X \left(1 + \int_0^t \psi(\omega) \, d\omega \right),$$

$$\frac{dy}{dt} = X \psi(t) ;$$

now when t is very great, $\frac{dy}{dt}$ is the steady flow of electricity through the glass divided by the capacity. Hence

$$\psi(\infty) = B. \quad\quad\quad\quad (3)$$

B is the reciprocal of the specific resistance multiplied by 4π and divided by the electrostatic capacity of the substance.

We have no practicable method of determining y_t; but we may proceed thus:—During insulation y_t is constant; we have then

$$x_t = A - \int_0^t x_{t-\omega} \, \psi(\omega) \, d\omega ; \quad\quad\quad\quad (4)$$

x_t and $x_{t-\omega}$ alone can be measured; (4) is, then, the equation by aid of which $\psi(\omega)$ must be determined.

* *Vide supra* p. 1.

(α) Let x_t be maintained constant $= X$ from time 0 to time t, then insulate ; differentiating (4),

$$\frac{dx_t}{dt} = - X\psi(t) - \int_0^t \frac{dx_{t-\omega}}{dt}\psi(\omega)\,d\omega$$
$$= - X\psi(t)$$
$$= - BX \text{ when } t \text{ is very great.}$$
$$\left.\begin{array}{c}\\\\\\\end{array}\right\} \dots\dots\dots\dots(5)$$

To find B, charge for a long time to a constant potential, insulate and instantly observe the rate of decrease of the potential.

(β) Let the flask be charged for a very short time τ and then be insulated; at the instant of insulation we have $\dfrac{dx_t}{dt} = - X\psi(\tau)$. Hence an approximation may be made to an inferior limit of $\psi(0)$.

(γ) Let x_t be constant $= X$ for a long time from $t = - T$ to $t = 0$; discharge and, after a further time t, insulate :—

$$x_{t+\tau} = A - X \int_{t+\tau}^{T+t+\tau} \psi(\omega)\,d\omega - \int_0^\tau x_{t-\omega+\tau}\,\psi(\omega)\,d\omega,$$
$$\frac{dx_t}{dt} = \{X\,\psi(t) - B\} \text{ when } \tau \text{ vanishes.}$$
$$\left.\begin{array}{c}\\\\\end{array}\right\} \dots\dots(6)$$

To find $\psi(t)$ in terms of t charge for a very long time, discharge and from time to time insulate and determine $\dfrac{dx_t}{dt}$.

(δ) Let the charging last during a shorter time τ', then discharge and insulate from time to time as in (γ) :—

$$\frac{dx_t}{dt} = X\{\psi(t) - \psi(\tau' + t)\}. \qquad \dots\dots\dots\dots\dots(7)$$

(ϵ) Charge during time τ', and reverse the charge for time τ'' before discharging :—

$$\frac{dx_t}{dt} = X\{\psi(t) - 2\psi(\tau'' + t) + \psi(\tau' + \tau'' + t)\}. \quad \dots\dots(8)$$

III. *Glass* No. 1.—This glass is a compound of silica, soda, and lime. In a damp atmosphere it "sweats," the surface showing a crystalline deposit easily wiped off. For a soda glass it is very white. Density 2·46.

When the flask was mounted, connected with the electrometer, the image from which was deflected 70 divisions by one Daniell's element, and insulated, it was found to steadily develop a negative charge, amounting to 11 scale-divisions in 10 seconds, and increasing to a maximum of 25 divisions. The cause of this the author cannot explain. Two other flasks of the same glass behaved in a similar manner—in one case, with the thin flask of § I., the charge rising to 40 divisions, with the thick flask to only 15 divisions. No sensible effect of the same kind was noticed with any other glass. The effect does not appear to be due to the connecting wires (for these were repeatedly removed and replaced by fresh ones), nor to difference between the acid within and that outside the flask, as this also was changed.

Experiment α.—The flask was charged to 500 divisions for half an hour, insulated, and the potential observed after 5, 10, 15, 20 seconds. The mean of several experiments gave for these times 372, 275, 216, 170 : hence the loss in 5 seconds is about 25 per cent.; and from this we may readily deduce $\frac{dx}{dt}$, since the percentage of loss is not materially different in the second interval of 5 seconds. $B = 3\cdot4$, the minute being unit of time.

Experiment β.—An attempt was made to estimate $\psi(0)$. The charging lasted one second. In two seconds from insulation the charge fell from 500 to about 330, which gives $\psi(0)$ certainly greater than 10 2. This can, of course, only be regarded as the roughest approximation.

Experiment γ.—The flask was charged positively for about 19 hours with 48 elements, the electromotive force of which is represented by about 3360 scale-divisions. It was then discharged, and at intervals insulated for 10 seconds, and the residual charge developed in that time observed. Column I. gives the time in minutes from first discharge to the middle of each 10-second period; II. the charge developed in 10 seconds; III. the estimated value of $\psi(t) - B$, obtained by correcting for the negative charge which it was found this flask took in 10 seconds, and dividing by 3360.

These results are certainly much below the true values, for the image moved over the scale much more rapidly in the first than in the second 5 seconds; but their ratios are probably fair approximations.

I.	II.	III.	I.	II.	III.
½	190	0·36	15	17	0·050
1	106	0·21	20	14	0·045
2	57	0·12	30	11	0·040
3	42	0·094	40	7	0·032
4	36	0·084	50	5	0·029
5	30	0·074	60	3	0·025
7	26	0·066	90	0	0·020
10	22	0·060	180	− 5	0·011

Experiment δ.—This experiment was tried both with a positive and a negative charge. The charge lasted 90 minutes. The readings were made as in γ.

I. gives the time in minutes;
II. the readings when the charge was positive;
III. when the charge was negative;
IV. the mean of II. and III.;
V. the value calculated from γ.

I.	II.	III.	IV.	V.
½	180	190	185	190
1	93	120	106	106
2	45	75	60	57
3	31	68	49	42
4	...	53	...	36
5	22	47	34	30
7	16	43	29	26

The same experiments were made, but with time of charging only 5 minutes.

Columns II. and III. give the means in each case of two separate observations, made on different occasions.

I.	II.	III.	IV.	V.
$\frac{1}{2}$	150	170	160	162
1	80	$92\frac{1}{2}$	$86\frac{1}{4}$	79
2	$22\frac{1}{2}$	$41\frac{1}{2}$	32	31
3	$10\frac{1}{2}$	29	$19\frac{3}{4}$	18
4	5	23	14	13
5	0	$20\frac{1}{2}$	$10\frac{1}{4}$	8
7	-4	$18\frac{1}{2}$	7	6

Glass No. 2.—This glass is of a deep blue colour; it is composed of silica, soda, and lime, the quantity of soda being less than in No. 1, but of lime greater. The colour is due to a small quantity of oxide of cobalt. The temperature throughout ranged from 62° F. to 64° F.

Experiment α.—The flask was charged for several minutes, and then insulated. The intensity of the charge before insulation, and at intervals of 5 seconds after, was observed, the whole experiment being repeated three times. The mean is given.

Time............... 0.   5.   10.   15.   20.   30.   40.
Reading......497   465   433·6   405   379   342   311
$$B = 0\cdot77.$$

Experiment β.—The charging lasted 2 seconds. The flask was then insulated, and its charge measured at intervals of 5 seconds. The mean of two fairly accordant observations is given.

Time .... 0.   5.   10.   15.   20.
Reading 490   390   325 $282\frac{1}{2}$ 249

Hence $\psi(0) > 2\cdot4$, probably much greater.

Experiment γ.—The flask was charged with 48 elements for 8 hours in the first experiment, and subsequently for 3 hours 25 minutes for a second experiment, the effect of the previous charging being still considerable when the charging began. After discharge the flask was from time to time insulated for 20 seconds, and the residual charge developed in that time was observed.

I. gives the time from discharge to the middle of the periods of insulation;

II. and III. the observations in the two experiments;

IV. the results corrected by a curve from II. and III.;

V. the values of $\psi(t) - B$, again not corrected for the rapid decrease in $\dfrac{dx}{dt}$ after each insulation.

It may be remarked that the image in this case moved in 10 seconds about $\frac{3}{5}$ of what it attained in 20 seconds.

I.	II.	III.	IV.	V.
$\frac{1}{2}$	470	468	469	0·42
1	300	325	310	0·28
2	178	183	180	0·16
3	134	134	133	0·12
4	105	107	106	0·094
5	89	91	90	0·080
7	68	69	68	0·061
10	52	54	53	0·047
15	39	41	40	0·036
20	32	36	34	0·030
30	...	29	28	0·025
60	...	20	20	0·018
90	16	...	16	0·014
600	...	6	6	0·005

Experiment δ.—The charging with 48 elements lasted 5 minutes. The experiment was tried twice with positive and negative charges respectively. II. and III. give the readings, whilst IV. gives the value calculated from the curve of γ.

I.	II.	III.	IV.
$\frac{1}{2}$	...	385	385
1	212	228	232
2	...	110	112
3	66	67	72
4	47	47	50
5	$34\frac{1}{2}$	$33\frac{1}{2}$	37
10	$11\frac{1}{2}$	10	13
15	5	$4\frac{1}{2}$	6

Experiment ε.—The flask was for many hours charged negatively, then positively for 5 minutes, and observations of residual charge were made as before.

Column III. are the values calculated from γ by the formula

$$\frac{dx}{dt} = \psi(t) - 2\psi(T + t).$$

I.	II.	III.
$\frac{1}{2}$	− 310	− 301
1	− 168	− 154
2	− 48	− 44
4	+ 8	+ 6
5	+ 17	+ 16
10	+ 28	+ 27
15	+ 27	+ 28
20	+ 27	+ 27

Glass No. 3.—Common window-glass, composed of silica, soda, and lime, the quantity of lime being greater than in No. 2. This glass does not "sweat" in a moist atmosphere. The temperature was 68° F.

Experiment α.—The flask was charged to 425 divisions for about 3½ hours, and was then insulated. After ½ minute the charge was 210; 1 minute, 138; 2 minutes, 74; 3 minutes, 50. Hence B is certainly greater than unity, and lies intermediate between the values for glasses 1 and 2.

Experiment γ gives the observed values of $\psi(t) - B$ throughout a little less than in No. 2. As this flask was not very well blown further experiments were not made.

If the values of $\psi(t) - B$ could be accurately obtained for these three glasses, they would certainly differ less from each other than they appear to do.

Glass No. 4.—Optical hard crown. Density 2·48. Composed of silica, potash, and lime. The composition may be regarded as corresponding to a glass intermediate between 1 and 3, with the soda replaced by potash.

The experiments α and β were made by the following modified method:—The whole battery of 48 elements was used, one pole being connected with the case of the electrometer and the exterior of the flask, the other with the interior of the flask by a cup of mercury and also with one electrode of the electrometer. The other electrode was permanently connected with the interior of the flask. It was ascertained that the image remained at zero whether both quadrants were charged equally or both discharged. The potential of the 48 elements was measured by 6 elements at a time; the extremes were 432 and 437, and the total 3475 scale-divisions. Where the charge of each quadrant is considerable and of the same sign, it cannot be assumed that the deflection for a given difference is the same as if the charges were small, or of equal and opposite sign; in fact, if the potentials of the quadrant and the jar of the electrometer are of the same sign, the sensibility of the instrument will be diminished (*vide* Maxwell's *Electricity and Magnetism*, vol. I. p. 273). On this account the results for $\psi(t)$ given below should be increased by about $\frac{1}{15}$ part of their value. The experiment consisted in insulating the flask from the battery, and observing the difference of potential between the flask and the battery after a suitable interval.

The flask was charged and instantly insulated at 8.25 P.M. The image traversed 164 divisions in 10 seconds. The flask was again connected with the battery, and insulated from time to time.

I. gives the middle of the period of insulation, measured from 8.25; II. the division traversed; III. the duration of insulation; IV. the value of $\psi(t)$.

I.	II.	III.	IV.
5 seconds.	164	10 seconds.	0·28
1 minute.	26	20 „	0·022
2 minutes.	14	20 „	0·012
3 „	11	20 „	0·0094
5 „	8	20 „	0·0069
10 „	34	2 minutes.	0·0049
15 „	28	2 „	0·0040
20 „	22	2 „	0·0031
30 „	36½	4 „	0·0026
60 „	25	4 „	0·0018
15 hours.	11	6 „	0·0005

Glass No. 5.—Optical soft crown. Density 2·55. Composed of silica and potash, with lead and lime in small quantity.

Experiments α and β.—68 divisions of the electrometer-scale equal one Daniell's element.

The flask was charged for 5 seconds, insulated, and the loss in the subsequent 10 seconds observed. The result may be regarded as giving an approximation to $\psi(\frac{1}{6})$. The mean of two experiments gives a fall from 471 to 452½, or $\psi(\frac{1}{6}) = 0·23$.

Charging for 45 seconds, and observing the loss during 30 seconds, gave $\psi(1) = 0·06$.

The flask was connected with the battery continuously, and only insulated at intervals, and connected with the electrometer for a short time to determine the rate of loss. The following values are thence deduced :—

t	5.	10.	30.	60.	120.	180.	300.
$\psi(t)$	0·025	0·017	0·012	0·009	0·007 +	0·007 −	0·006

$\psi(\infty)$ probably does not differ much from 0·005 or 0·004.

Experiment γ.—The flask was charged for 3 days with 48 elements, equal to 3260 divisions, or thereabouts, then discharged.

I. gives the time from first discharge to the middle of the period of insulation;

II. the scale-divisions traversed;

III. the times of insulation in minutes;

IV. the value of $\dfrac{dx_t}{dt}$;

V. $\psi(t) - B$.

I.	II.	III.	IV.	V.
$\frac{1}{2}$	53	$\frac{1}{6}$	318	0·098
1	62	$\frac{1}{3}$	186	0·057
2	64	$\frac{1}{2}$	128	0·039
3	62	$\frac{2}{3}$	93	0·029
5	70	1	70	0·021
10	92	2	46	0·014
15	71$\frac{1}{2}$	2	35·75	0·011
20	63	2	31·5	0·0097
30	48$\frac{1}{2}$	2	24·25	0·0074
60	109	8	13·6	0·0042
90	89	8	11·12	0·0034
125	69$\frac{1}{2}$	8	8·7	0·0027
180	54	8	6·75	0·0021

The results thus obtained agree fairly with those obtained by Experiment β; the differences may be attributed to errors of observation.

Experiment δ.—The charging lasted 5 minutes. The experiment was performed twice, with positive and negative charges respectively.

I. gives the time from first discharge;

II. the period of insulation;

III. and IV. the divisions traversed in that time;

V. their mean;

VI. the value of $\dfrac{dx_t}{dt}$ thence obtained;

VII. the value of $\dfrac{dx_t}{dt}$ calculated from the last experiment.

I.	II.	III.	IV.	V.	VI.	VII.
$\frac{1}{2}$	$\frac{1}{6}$	37	...	37	222	252
1	$\frac{1}{3}$	42	43	$42\frac{1}{2}$	127·5	124
2	$\frac{1}{2}$	32	$32\frac{1}{2}$	$32\frac{1}{4}$	64·5	72
5	1	23	$22\frac{1}{2}$	$22\frac{3}{4}$	22·75	24
15	4	23	...	23	5·75	4·25

The differences between VI. and VII. are somewhat large; they may perhaps be in part attributed to the fact that $\dfrac{dx_t}{dt}$ is deduced from observations on a quantity not uniformly increasing, on the assumption that the increase is uniform, and to the inequality of the times of insulation.

Glass No. 6.—A flint glass containing less lead than No. 7.

Experiments α *and* β.—66 divisions of the scale equal to one Daniell's element.

The flask was continuously connected with the battery, and only insulated for brief periods, to determine the rate of loss, the following values are thence deduced:—

t	1.	5.	15.	120.	240.
$\psi(t)$	0·013	0·007	0·004	0·0016	0·001

Experiment γ.—The flask was charged for 13 hours with 48 elements, then discharged.

The columns are the same as in glass No. 5.

I.	II.	III.	IV.	V.
1	21	$\frac{1}{3}$	63	0·02
5	37$\frac{1}{2}$	2	18·75	0·006
15	48	6	8·0	0·0026
75	60	24	2·5	0·0008

There is a considerable discrepancy between the values of $\psi(1)$ from α and $\psi(1) - B$ from γ; the former may be in error, as it was deduced from the time of traversing 3 divisions only.

Glass No. 7.—Optical "light flint." Density 3·2. Composed of silica, potash, and lead. Almost colourless. The surface neither "sweats" nor tarnishes in the slightest degree. This glass at ordinary temperatures is sensibly a perfect insulator.

A flask was mounted in the usual way on July 15th; it was charged with 48 elements for some hours, the potential being 240 scale-divisions as measured through the "induction-plate" of the electrometer. The charging-wire was then withdrawn. On July 23rd the wire was again introduced and connected with the induction-plate; a charge of 183 scale-divisions still remained, although the temperature of the room was as high as 72° F. The flask was again put away till Aug. 9th, when the charge was found to be 178. On September 14th it was 163. Lastly on October 14th it had fallen to 140.

As might be expected from the last experiment, the residual charge in this glass is small. The flask was charged for 9 hours with 48 elements; it was discharged, and after 4 minutes insulated; in 2 minutes the residual charge had only attained 11$\frac{1}{2}$ divisions, giving $\psi(5) = 0\cdot0017$. It was again insulated after 44 minutes; in 12 minutes the charge was 10$\frac{1}{2}$, giving $\psi(50) = 0\cdot00026$.

Since the loss by conduction is so small, the flask may be strongly charged by an electrophorus instead of with the battery. If it is left insulated for a considerable time, and then discharged, and the return charge observed, it may be assumed that the exciting charge has been sensibly constant during the latter portion of the period of insulation.

The flask was strongly charged and remained insulated for 3 hours 40 minutes; it was then discharged, and from time to time was temporarily insulated to ascertain the rate of return of charge.

At $\frac{1}{2}$ minute   250 divisions in $\frac{1}{6}$ minute $= 1500$ per minute.

5	minutes	247	„	1	„	=	247	„	„
10	„	285	„	2	minutes	=	$142\frac{1}{2}$	„	„
15	„	304	„	3	„	=	101	„	„
30	„	326	„	6	„	=	54	„	„

It was immediately charged again, insulated for 70 minutes, and then the observations repeated.

At $\frac{1}{2}$ minute   120 divisions in $\frac{1}{6}$ minute $= 720$ per minute.

1	„	135	„	$\frac{1}{3}$	„	= 405	„	„
2	minutes	125	„	$\frac{1}{2}$	„	= 250	„	„
5	„	121	„	1	„	= 121	„	„
10	„	142	„	2 minutes	= 71	„	„
15	„	106	„	2	„	= 53	„	„

The ratios of the numbers in the two experiments agree fairly.

Glass No. 8.—"Dense flint." Density 3·66. Composed of silica, lead, and potash, the proportion of lead being greater than in No. 7.

Experiment α.—The flask was charged for 3 hours to 500 divisions, and then insulated :—

After 1 minute from insulation 499$\frac{3}{4}$

„ 5 minutes „   „ 499

„ 30 „   „   „ 495

Hence $\psi(180) = 0.0004$.

Experiment β.—The flask was charged for 5 seconds, insulated, and the potential read off at intervals of ½ minute. The results are the mean of two observations :—

Reading ... 497   479½   475½   474   473   472½   472¼
Time 0   ½   1   1½   2   2½   3

From this it may be inferred that $\psi(0)$ is considerably greater than 0·07. An experiment on residual charge gives

$$\psi(1) - B = 0\cdot017.$$

Glass No. 9.—Extra dense flint. Density 3·88. Colour slightly yellow The proportion of lead is somewhat greater than in No. 8. The surface tarnishes slowly if exposed unprotected to the air.

The flask was charged for 10 seconds to 500, and was then insulated.

After   1 minute   the reading was 499
  „   3 minutes „   „   „   497¾
  „   5   „   „   „   „   495
  „   30   „   „   „   „   486½
  „   60   „   „   „   „   479

The flask was charged with 48 elements for 1½ hour, and the residual charge observed,

$$\psi(2) - B = 0\cdot003.$$

An attempt was made to obtain a knowledge of the form of the function $\psi(t)$ in the same manner as for No. 7. The flask was charged from the electrophorus, and allowed to stand insulated for 22 hours; it was then discharged and temporarily insulated at intervals.

At   ½ min. traversed 130 divisions in   ⅙ min. = 780 per min.
 „   1   „   „   160   „   „   ⅓ „ = 480   „
 „   2   „   „   145   „   „   ½ „ = 290   „
 „   5   „   „   152   „   „   1 „ = 152   „
 „   10   „   „   189   „   „   2 „ = 94½ „
 „   15   „   „   217   „   „   3 „ = 72   „
 „   30   „   „   275   „   „   6 „ = 46   „
 „   60   „   „   360   „   „ 12 „ = 30   „
 „ 120   „   „   437   „   „ 24 „ = 18   „

It will be remarked that in this case $\psi(t) - B$ deviates further from the reciprocal of the time than in the case of No. 7.

Glass No. 10.—Opal glass. This glass is white and opaque. It is essentially a flint. The reason for examining it was to ascertain if its opacity had any striking effect on its electrical properties.

Experiment α.—The flask was charged to 462 divisions for 5 hours; on insulation the loss was found to be 4 to 5 divisions in an hour; hence $B = 0\cdot00016$.

Experiment β.—Charged to 462 for 10 seconds; a loss of 2 in 3 minutes was observed on insulation.

Experiment γ.—The flask was charged with 48 elements, each equal to 67 divisions of the scale, for 5 hours, and was then discharged.

At 1   minute, 4½ divisions in   ⅓ minute,

„ 2½ minutes, 6   „   „ 1   „

„ 5   „   6   „   „ 2 minutes,

or

$$\psi(1) - B = 0\cdot004,$$
$$\psi(2\tfrac{1}{2}) - B = 0\cdot002,$$
$$\psi(5) - B = 0\cdot001.$$

The residual charge is smaller than in any other glass observed.

A few of the results of the preceding experiments are collected in the following Table for the purpose of ready comparison.

I.   The greatest value of ψt observed.

II.   „ least   „   „   „

III. $\psi(1) - B$ as obtained by experiment γ.

IV. $\psi(5) - B$ „   „   „

V. $\psi(60) - B$ „   „   „

Glass	I.	II.	III.	IV.	V.
1	10·2	3·4	0·21	0·073	0·025
2	2·45	0·76	0·28	0·08	0·018
3	....	1·0	....	0·05	0·01
4	0·23	0·0005	0·0215	0·0064	0·0013
5	0·23	0·006	0·057	0·021	0·0042
6	0·013	0·001	0·02	0·006	1·0008*
7	....	0·00002	....	0·0017	0·00026†
8	0·07	0·0004	0·017	....	....
9	0·002	....	0·003‡	....	....
10	0·0014	0·00016	0·004	0·001	....

From this Table two classes can at once be selected as having
well-marked characters.   The soda-lime glasses, although the
composition and colour vary widely, agree in possessing small
insulating power, but exhibit very great return charge.   The
values of the function $\psi(t) - B$ for the three glasses agree almost
within the limits of these roughly approximate experiments.

At the opposite extreme are the flints or potash-lead glasses,
which have great specific resistance.   The experiment does not
prove that No. 7 conducts electricity at all; for it is not certain
that the very slight loss of charge may not be due to conduction
over the surface of the glass; but it is certainly not less than
100,000 times as resistant as No. 1.   The flints also have very
similar values of $\psi(t) - B$, much smaller than the soda-lime
glasses.

IV.   It is known that glass at a moderately high temperature
conducts electricity electrolytically.   The following experiment

* $\psi(75) - B$.   † $\psi(50) - B$.   ‡ $\psi(20) - B$.

shows that with the more conductive glasses electrolytic conduction
occurs at the ordinary temperature of the air.

A flask of blue glass, No. 2, was very carefully insulated with
strong sulphuric acid within the flask, and was placed in a vessel
of caustic potash. Platinum wires dipping in the two liquids
communicated with the quadrants of the electrometer. On
insulation the acid developed a positive charge as follows :—

In $\frac{1}{2}$ minute 15 divisions of the scale,
„ 1   „ $22\frac{1}{2}$ „   „   „
„ 2 minutes $33\frac{1}{2}$ „   „   „
„ 5   „   47   „   „   „
„ 10   „   55   „   „   „
„ 15   „   57   „   „   „

one Daniell's element giving 68 divisions of the scale.

The experiment was repeated after the flask had stood some
days with the two liquids connected by a platinum wire ; the
potential developed much more slowly, and in 50 minutes was
stationary at $38\frac{1}{2}$ divisions.

Summary.—These experiments are subject to many causes
of error. Deducing $\dfrac{dx_t}{dt}$ from an observation of dx_t in a period
of many seconds or even minutes gives values of $\psi(t) - B$ necessarily
too low, in some cases very much too low. No attempt was made
to keep the glass at a constant temperature ; the temperature of
the room was occasionally noted, but is not given here, as no
conclusion is based upon it. The experiments were performed
irregularly at such times as other circumstances permitted. It
will be observed that the discords of the experiments of verification
are considerable, but they are irregular. It may, perhaps, be
assumed that they are within the limits of error, and we may
infer that the fundamental hypothesis is verified, viz. that the
effects on a dielectric of past and present electromotive forces are
superposable. Ohm's law asserts the principle of superposition
in bodies in which conduction is not complicated by residual
charge. Conduction and residual charge may be treated as parts
of the same phenomenon, viz. an after-effect, as regards electric
displacement, of electromotive force. The experiments appear to
show, though very roughly, that the principle of Ohm's law is
applicable to the *whole* phenomenon of conduction through glass.

V.   Effect of Temperature.

The purpose of the previous experiments being to examine generally the applicability of the formulæ and to compare the values of $\psi(t)$ for different glasses of known composition, no account was taken of temperature, and no attempt made to maintain it constant, although it is well known that changes of temperature greatly affect both conduction and polarization in glass*. It appeared, however, desirable to compare the same glass at different temperatures in the same manner as different glasses at the same temperature.

The flask, carefully filled with sulphuric acid as before, was placed in an earthenware jar containing sulphuric acid, which was in its turn placed in a double cylindrical shell of copper, with oil or water between the cylinders. The jar was covered by two disks of wood, through holes in the centre of which the neck of the flask projected. A cap of sealing-wax, carrying a small cup of mercury for making electrical connexions with the interior, closed the flask. A thermometer dipped into the acid outside the flask for reading the temperature of the glass, whilst a second thermometer was inserted between the cylinders in the oil or water to help the observer in regulating the temperature by means of a spirit-lamp. In the two experiments below freezing-point the earthenware jar was removed from the oil-bath and placed in a freezing-mixture of hydrochloric acid and sulphate of soda. In all cases the temperature was maintained approximately constant for some time before observing. It will be remarked that, as the acid was not stirred, the temperature-readings are subject to a greater probable error than that due to the thermo-meter itself. But as the changes of temperature of the acid were always very slow, the error thus introduced cannot seriously affect the results. All temperatures are Centigrade. The actual readings are given, and also the temperature, roughly corrected when necessary, for the exposed portion of the stem of the thermometer. The times in these and in most of the previous experiments were taken by ear from a dead-beat seconds clock, the eye being fixed on the image and the scale. In the intervals between the short insulations to determine $\dfrac{dx}{dt}$, the flask was either connected with the battery or discharged. In all cases the

* *Vide* Mr Perry, *Proceedings of the Royal Society*, 1875, p. 468; Prof. Clerk Maxwell, *Electricity and Magnetism*, Art. 271.

registered time of observation is taken at the middle of the
period of insulation; thus, in the experiment at $39\frac{1}{2}°$ below,
insulation was made 1 second before the minute, and the reading
1 second after.   Two glasses were examined, Nos. 2 and 7, selected
as extreme cases.   The whole of the observations made are given,
excepting three manifestly in error, although only a portion are
used.   The values of $\psi(5)$ and $\psi(10)$, for glasses 2 and 7
respectively, are taken as sensibly equal to B, and are calculated
on the assumption that during the short time of insulation the
rate of loss at any instant is proportional to the then charge*.

The values of $\psi(1) - B$ and $\psi(5) - B$ are deduced as though $\dfrac{dx}{dt}$
were constant during the time of insulation, and are therefore
considerably below the truth in all cases.   It will be observed that
the battery was not quite constant; but the value of 48 elements
may be taken as 3160 scale-divisions without serious error.

Glass No. 2.—Temperature 53°.   It was roughly estimated
that on insulation $\frac{1}{4}$ of the charge was lost within 1 second.
Notwithstanding this high conductivity, the residual charge was
capable of rising to more than 400 scale-divisions when the flask
had been charged with 48 elements and then discharged for a
few seconds.   This differentiates the polarization in even highly
conductive glass from the electrochemical polarization in a voltameter, in a single element of which no electromotive force can
give rise to a return force greater than that due to the energy of
combination of the constituents of the electrolyte.   Subsequently,
considerable residual charges were obtained with the same glass
up to 150°; at 180° the residual charge was so rapidly lost that
it was hardly sensible.

Temperature $39\frac{1}{2}°$.

h. m.

Time 6 10.   Charged with 7 elements.

6 11.   From 462 to 350 in 2 seconds.

6 12.   „   463 to 350   „   „

6 17.   „   464 to 350   „   „

6 19.   „   464 to 350   „   „

$$\left. \begin{matrix} B = 8.46 \\ \text{Log } B = 9.27 \end{matrix} \right\} \text{ at } 39\tfrac{1}{2}°.$$

---

* In the original paper as published in the *Phil. Trans.* there was an error in
this calculation, which resulted in the values of B being all too high in the same
proportion—about 15 per cent.   It probably arose from the use of a wrong value of
the constant $\log_e 10$.   In this reprint it is corrected. [ED.]

h m
Time 6 20. Charged with 48 elements.
           Temperature 41°.
  6 40. Discharge.
  6 41. 50 in 4 seconds.
  6 42. 28   ,,   ,,
  6 43. 18   ,,   ,,
           Temperature 41°.

$$\psi(1) - B = 0{\cdot}24 \text{ at } 41°.$$

Temperature $33\frac{1}{2}°$.
7 50. Charged with 7 elements.
7 51. 462 to 340 in 4 seconds.
7 52. 463 to 340 ,,   ,,
7 55. 465 to 343 ,,   ,,
      Temperature $33\frac{1}{4}°$.

$$\left.\begin{array}{l} B = 4{\cdot}6 \\ \text{Log } B = 0{\cdot}66 \end{array}\right\} \text{ at } 33\frac{3}{8}°.$$

7 56. Charged with 48 elements.
      Temperature 35°.
8 30. Discharge.
8 31. 115 in 10 seconds.
8 32. 67   ,,   ,,
8 33. 46   ,,   ,,
8 35. 29   ,,   ,,

$$\left.\begin{array}{l} \psi(1) - B = 0{\cdot}22 \\ \psi(5) - B = 0{\cdot}055 \end{array}\right\} \text{ at } 35°.$$

Temperature $27\frac{1}{2}°$.
10 2. Charged with 7 elements.
10 3. 459 to 340 in 5 seconds.
10 4. 460 to 360 ,,   ,,
10 7. 461 to 368 ,,   ,,
      Temperature 27°.

$$\left.\begin{array}{l} B = 2{\cdot}7 \\ \text{Log } B = 0{\cdot}43 \end{array}\right\} \text{ at } 27\frac{1}{4}°.$$

h m

Time 10 8. Charged with 48 elements.

            Temperature 28°.

10 41. Discharge.

10 42. 140 in 10 seconds.

10 43. 77   ,,   ,,

10 44. 53   ,,   ,,

10 46. 34   ,,   ,,

$$\left.\begin{array}{l}\psi(1)-B=0\cdot26\\\psi(5)-B=0\cdot064\end{array}\right\} \text{ at } 28°.$$

Temperature 26°.

8 45. Charged with 7 elements.

8 46. 452 to 350 in 6 seconds.

8 47. 453 to 350   ,,   ,,

8 50. 455 to 368   ,,   ,,

         Temperature $25\frac{1}{2}°$.

$$\left.\begin{array}{l}B=2\cdot1\\\text{Log } B=0\cdot32\end{array}\right\} \text{ at } 25\frac{3}{4}°.$$

Temperature $24\frac{1}{4}°$.

9 11. Charged with 7 elements.

9 12. 458 to 345 in 8 seconds.

9 13. 458 to 351   ,,   ,,

9 16. 457 to 355   ,,   ,,

         Temperature 24°.

$$\left.\begin{array}{l}B=1\cdot9\\\text{Log } B=0\cdot28\end{array}\right\} \text{ at } 24\frac{1}{8}°.$$

Temperature $22\frac{1}{2}°$.

9 38. Charged with 7 elements.

9 39. 455 to 338 in 10 seconds.

9 40. 456 to 340   ,,   ,,

9 43. 457 to 352   ,,   ,,

         Temperature $22\frac{1}{4}°$.

$$\left.\begin{array}{l}B=1\cdot56\\\text{Log } B=0\cdot19\end{array}\right\} \text{ at } 22\frac{3}{8}°.$$

h m

Time 9 45. Charged with 48 elements.

Temperature $20\frac{1}{2}°$.

10 15. Discharged.

10 16. 150 in 10 seconds.

10 17.   81   „   „

10 18.   55   „   „

10 20.   33   „   „

$$\left.\begin{array}{l} \psi(1) - B = 0\cdot28 \\ \psi(5) - B = 0\cdot062 \end{array}\right\} \text{ at } 20\frac{1}{2}°.$$

Temperature $7\frac{1}{4}°$.

4 40. Charged with 7 elements.

4 41. 466 to 385 in 20 seconds.

4 42. 465 to 397   „   „

4 45. 466 to 411   „   „

$$\left.\begin{array}{l} B = 0\cdot37 \\ \text{Log } B = \bar{1}\cdot57 \end{array}\right\} \text{ at } 7\frac{1}{4}°.$$

4 46. Charged with 48 elements.

Temperature $7\frac{1}{4}°$.

5 15. Discharge.

5 16. 250 in 20 seconds.

5 17. 160   „   „

5 18. 110   „   „

5 20.   66   „   „

$$\left.\begin{array}{l} \psi(1) - B = 0\cdot24 \\ \psi(5) - B = 0\cdot062 \end{array}\right\} \text{ at } 7\frac{1}{4}°.$$

Temperature $-3°$, after standing 30
minutes in the freezing-mixture.

7 19. Charged with 7 elements.

7 20. 457 to 417 in 20 seconds.

7 21. 458 to 427   „   „

7 24. 459 to 438   „   „

7 29. 461 to 442   „   „

Temperature $-3°$.

$$\left.\begin{array}{l} B = 0\cdot14 \\ \text{Log } B = \bar{1}\cdot15 \end{array}\right\} \text{ at } -3°.$$

h m

Time 7 30. Charged with 48 elements.

Temperature $-1\frac{1}{4}°$.

8   3. Discharged.

8   4. 180 in 20 seconds.

8   5. 115   „   „

8   6. 83   „   „

8   8. 56   „   „

Temperature $-1°$.

$$\left.\begin{array}{l}\psi(1) - B = 0\cdot17 \\ \psi(5) - B = 0\cdot053\end{array}\right\} \text{ at } -1\tfrac{1}{8}°.$$

Temperature $-5°$, in a fresh freezing-mixture.

8 48. Charged with 7 elements.

8 49. 463 to 432 in 20 seconds.

8 50. 464 to 438   „   „

8 53. 465 to 447   „   „

$$\left.\begin{array}{l}B = 0\cdot12 \\ \text{Log } B = \overline{1}\cdot08\end{array}\right\} \text{ at } -5°.$$

8 55. Charged with 48 elements.

Temperature $-3°$.

9 25. Discharged.

9 26. 176 in 20 seconds.

9 27. 108   „   „

9 28. 80   „   „

9 30. 53   „   „

$$\left.\begin{array}{l}\psi(1) - B = 0\cdot17 \\ \psi(5) - B = 0\cdot050\end{array}\right\} \text{ at } -3°.$$

As in Mr Perry's experiments the results agree closely with the formula

$$\text{Log } B = a + b\theta,$$

where θ is the temperature, and in this case $a = \bar{1}\cdot 28$ and $b = 0\cdot 0415$. The following Table gives the observed and calculated values:—

Temp.	B observed.	B from formulæ.
$39\frac{1}{2}^{\circ}$	8·4	8·2
$33\frac{3}{8}$	4·6	4·7
$27\frac{1}{4}$	2·7	2·5
$25\frac{3}{4}$	2·12	2·2
$24\frac{1}{4}$	1·9	1·9
$22\frac{3}{8}$	1·56	1·6
$7\frac{1}{8}$	0·37	0·40
-3	0·14	0·15
-5	0·12	0·12

The residual charge results do not show so great a degree of regularity, probably because the direct deduction of $\dfrac{dx}{dt}$ as equal to $\dfrac{\delta x}{\delta t}$ gives a greater error than the method used for obtaining B. This much is quite certain, that the value of $\psi(1) - B$ and $\psi(5) - B$ is rapidly increasing up to $7°$. It appears probable that at higher temperatures these do not increase so rapidly if at all; but this is by no means certain, as although shorter times of insulation were used, the values at higher temperatures are notwithstanding more reduced by conduction than at the lower.

Glass No. 7.—Temperature 119°.

h m
Time 6 21. Charged with 7 elements.
6 22. 463 to 390 in 20 seconds.
6 23. 464 to 399 „   „
6 26. 465 to 412 „   „
6 31. 465 to 419 „   „
Temperature 119°.

$$B = 0\cdot 31$$
$$\text{Log } B = \bar{1}\cdot 49 \quad \text{at } 120\frac{1}{4}°.$$
$$\psi(1) = 0\cdot 511$$

h   m
Time 6 32.   Charged with 48 elements.
    Temperature 122°.
7   5.   Discharged.
7   6.   226 in 20 seconds.
7   7.   141 „ „
7   8.   104 „ „
7 10. 65 „ „

$$\left.\begin{array}{l} \psi\,(1) - B = 0\cdot215 \\ \psi\,(5) - B = 0\cdot062 \end{array}\right\} \text{ at } 123\tfrac{1}{4}°.$$

Temperature 107°.
7 51.   Charged with 7 elements.
7 53.   466 to 437 ? in 20 seconds.
7 56.   466 to 429 „ „
8   1.   466 to 447 „ „
    Temperature 107°.

$$\left.\begin{array}{l} B = 0\cdot126 \\ \text{Log } B = \bar{1}\cdot099 \end{array}\right\} \text{ at } 108°.$$

8   2.   Charged with 48 elements.
    Temperature 107°.
8 36.   Discharged.
8 37.   162 in 20 seconds.
8 38.   100 „ „
8 39. 76 „ „
8 41. 51 „ „

$$\left.\begin{array}{l} \psi\,(1) - B = 0\cdot155 \\ \psi\,(5) - B = 0\cdot05 \end{array}\right\} \text{ at } 108°.$$

9 25.   Charged with 48 elements.
    Temperature 98°.
10   1.   Discharged.
10   2.   110 in 20 seconds.
10   3. 74 „ „
10   4. 56 „ „
10   6. 39 „ „
    Temperature 97$\tfrac{3}{4}$°.

$$\left.\begin{array}{l} \psi\,(1) - B = 0\cdot11 \\ \psi\,(5) - B = 0\cdot037 \end{array}\right\} \text{ at } 98\tfrac{3}{4}°.$$

h m

Temperature $172\frac{1}{2}°$.

Time 7 25. Charged with 7 elements.

7 26. 461 to 270 in 3 seconds.

7 27. 462 to 272 ,, ,,

7 30. 463 to 277 ,, ,,

7 35. 465 to 281 ,, ,,

Temperature $172°$.

$$\left.\begin{array}{l} B = 10\cdot1 \\ \text{Log } B = \ 1\cdot006 \end{array}\right\} \text{at } 175\frac{1}{2}°.$$

7 36. Charged with 48 elements.

Temperature $172°$.

7 50. Discharged.

7 51. 100 in 5 seconds.

7 52.   50 ,, ,,

7 53.   28 ,, ,,

7 55.   9 ,, ,,

Temperature $171\frac{1}{2}°$.

$$\left.\begin{array}{l} \psi(1) - B = 0\cdot38 \\ \psi(5) - B = 0\cdot034 \end{array}\right\} \text{at } 175°.$$

9 0. Charged with 48 elements.

Temperature $150°$.

9 30. Discharged.

9 31. 122 in 5 seconds.

9 32. 125 in 10 seconds.

9 33.   96 ,, ,,

9 35.   64 ,, ,,

$$\left.\begin{array}{l} \psi(1) - B = 0\cdot46 \\ \psi(5) - B = 0\cdot12 \end{array}\right\} \text{at } 152\frac{1}{4}°.$$

Temperature $162°$.

10 13. Charged with 7 elements.

10 14. 461 to 330 in 3 seconds.

10 15. 462 to 340 ,, ,,

10 18. 463 to 346 ,, ,,

10 23. 463 to 353 ,, ,,

Temperature $161°$.

$$\left.\begin{array}{l} B = 5\cdot46 \\ \text{Log } B = 0\cdot737 \\ \psi(1) = 6\cdot66 \end{array}\right\} \text{at } 164°.$$

h m

Time 10 24.   Charged with 48 elements.

Temperature 165°.

10 58.   Discharged.

10 59.   125 in 5 seconds.

11 0.   74   „   „

11 1.   50   „   „

11 3.   26   „   „

$$\left.\begin{array}{l} \psi\,(1) - B = 0{\cdot}47 \\ \psi\,(5) - B = 0{\cdot}098 \end{array}\right\} \text{ at } 167\tfrac{1}{2}°.$$

Temperature 143°.

4 48.   Charged with 7 elements.

4 49.   469 to 400 in 4 seconds.

4 50.   469 to 403   „   „

4 53.   470 to 410   „   „

4 58.   470 to 420   „   „

Temperature $143\tfrac{1}{2}°$.

$$\left.\begin{array}{l} B = 1{\cdot}69 \\ \text{Log } B = 0{\cdot}228 \\ \psi\,(1) = 2{\cdot}38 \end{array}\right\} \text{ at } 145\tfrac{1}{4}°.$$

5 0.   Charged with 48 elements.

Temperature 143°.

5 23.   Discharged.

5 24.   190 in 10 seconds.

5 25.   115   „   „

5 26.   88   „   „

5 28.   55   „   „

$$\left.\begin{array}{l} \psi\,(1) - B = 0{\cdot}36 \\ \psi\,(5) - B = 0{\cdot}105 \end{array}\right\} \text{ at } 144\tfrac{3}{4}°.$$

Temperature 127°.

7 8.   Charged with 7 elements.

7 9.   465 to 412 in 10 seconds.

7 10.   466 to 416   „   „

7 13.   467 to 427   „   „

7 18.   468 to 428   „   „

$$\left.\begin{array}{l} B = 0{\cdot}54 \\ \text{Log } B = \bar{1}{\cdot}73 \\ \psi\,(1) = 0{\cdot}74 \end{array}\right\} \text{ at } 128\tfrac{1}{2}°.$$

h m
Time 7 20.   Charged with 48 elements.
              Temperature 126°.
7 58.   Discharged.
7 59.   135 in 10 seconds.
8 0.   86   „   „
8 1.   73   „   „
8 3.   47   „   „

$$\left.\begin{array}{l} \psi\,(1) - B = 0\cdot26 \\ \psi\,(5) - B = 0\cdot09 \end{array}\right\} \text{ at } 127\tfrac{1}{2}°.$$

              Temperature 79°.
9 30.   Charged with 7 elements.
9 35.   468 to 448 in 2 minutes.
9 40.   468 to 450   „   „
              Temperature 79°.

$$\left.\begin{array}{l} B = 0\cdot019 \\ \text{Log } B = \bar{1}\cdot28 \end{array}\right\} \text{ at } 79\tfrac{1}{2}°.$$

5 15.   Charged with 48 elements.
              Temperature 66°.
5 45.   Discharged.
5 46.   55 in 40 seconds.
5 47.   31   „   „
5 48.   22   „   „
5 50.   14   „   „
              Temperature 64½°.

$$\left.\begin{array}{l} \psi\,(1) - B = 0\cdot026 \\ \psi\,(5) - B = 0\cdot007 \end{array}\right\} \text{ at } 65\tfrac{1}{2}°.$$

              Temperature 94°.
6 35.   Charged with 7 elements.
6 37.   457 to 422 in 1 minute.
6 40.   458 to 432   „   „
6 47.   458 to 433   „   „
              Temperature 94½°.

$$\left.\begin{array}{l} B = 0\cdot057 \\ \text{Log } B = \bar{2}\cdot75 \end{array}\right\} \text{ at } 95°.$$

h   m

Temperature $153\frac{1}{2}°$.

Time 8   0.   Charged with 7 elements.

8   1.   461 to 340 in 4 seconds.

8   2.   461 to 350   ,,   ,,

8   3.   462 to 352   ,,   ,,

8   5.   463 to 358   ,,   ,,

8 10.   463 to 362   ,,   ,,

Temperature $153\frac{1}{2}°$.

$$
\left.\begin{array}{l} B = 3\cdot69 \\ \text{Log } B = 0\cdot57 \\ \psi\,(1) = 4\cdot54 \end{array}\right\} \text{ at } 155\frac{3}{4}°.
$$

Temperature 66°.

10 31.   Charged with 7 elements.

10 41. $464\frac{1}{2}$ to 445 in 4 minutes.

Temperature 67°.

$$
\left.\begin{array}{l} B = 0\cdot0105 \\ \text{Log } B = \bar{2}\cdot021 \end{array}\right\} \text{ at } 66\frac{3}{4}°.
$$

With this glass the results do not agree so closely with the exponential formula as with glass No. 2. This is perhaps not surprising when it is considered that the temperatures differ more from that of the room, and, consequently, that errors due to unequal heating of the acid, and to exposure of the stem of the thermometer, will be greater.

The observed values of B, and those calculated from the formula $\log B = \bar{4}\cdot10 + 0\cdot0283\theta$, are given in the following Table:—

θ.	Observed.	Calculated.
$175\frac{1}{2}°$	10·1	12
164	5·46	5·5
$155\frac{3}{4}$	3·69	3·2
$145\frac{1}{8}$	1·69	1·6
$128\frac{1}{2}$	0·54	0·54
$120\frac{1}{4}$	0·31	0·31
108	0·126	0·14
95	0·057	0·062
$79\frac{1}{2}$	0·019	0·022
66	0·010	0·009

The values obtained for $\psi(1)$ and B do not in general give a value of $\psi(1) - B$, which agrees very closely with that obtained by residual charge. This is not astonishing, for $\psi(1)$ and B are both subject to a considerable probable error, and do not differ greatly from each other. On the other hand, at high temperatures, the values of $\psi(1) - B$ and $\psi(5) - B$, obtained by residual charge, are undoubtedly much too low. It is interesting to remark, that whereas the values of $\psi(1) - B$ and $\psi(5) - B$ from residual charge do not increase with temperature above 160°, the values of $\psi(1) - B$ obtained by difference show a continually accelerated increase. The observed values of $\psi(1) - B$ and $\psi(5) - B$ are collected in the following Table. The values above 140°, if admitted at all, must be regarded as subject to an enormous probable error.

Temperature.	$\psi(1) - B$.	$\psi(5) - B$.
175°	0·38	0·034
167½	0·47	0·098
152¼	0·46	0·12
144¾	0·36	0·105
127½	0·26	0·09
123¼	0·215	0·062
108	0·155	0·05
98¾	0·11	0·037
65¼	0·026	0·007

It should be mentioned that the temperature experiments were not made on the same flask as flask No. 7 of the previous experiments, but on a flask of the same composition.

20.

REFRACTIVE INDICES OF GLASS.

[From the *Proceedings of the Royal Society*, No. 182, pp. 1—8, 1877.]

Most of the following determinations were made two years ago. They were not published at once, because the results showed more variation than was expected. They are now made known for two reasons. First, most of the glasses examined are articles of commerce, and can be readily obtained by any person experimenting upon the physical properties of glass; these glasses only vary within narrow limits, and their variations may be approximately allowed for by a knowledge of their density. Second, most of the prisms having three angles from each of which determinations were made, the probable error of the mean is very small, and any error of the nature of a blunder is certainly detected.

The form in which to present these results was a matter of much consideration. A curve giving the refractive indices directly is unsuitable, for the errors of observation are less than the errors of curve-drawing would be. The theory of dispersion is not in a position to furnish a satisfactory rational formula. The most frequently used empirical formula is $\mu = a + b\dfrac{1}{\lambda^2} + c\dfrac{1}{\lambda^4} + \ldots$, where λ is the wave-length of the ray to which μ refers. But to bring this within errors of observation it is necessary to include $\dfrac{1}{\lambda^6}$, which appears to be almost as important a term as $\dfrac{1}{\lambda^4}$. There are two points of importance in the selection of an empirical form : first, it must accurately represent the facts with the use of the fewest arbitrary parameters; second, it must be practically convenient for the purposes for which the results are useful.

In the present case the most convenient form is

$$\mu - 1 = a \{1 + bx(1 + cx)\},$$

where x is a numerical name for the definite ray of which μ is the refractive index. In the present paper line F, being intermediate between the strongest luminous and chemical rays, is taken as zero. Four glasses, Hard Crown, Soft Crown, Light Flint, and Dense Flint, are selected on account of the good accord of the results, and the mean of their refractive indices $\bar{\mu}$ is ascertained for each ray ; this is taken as a standard scale in which $x = \bar{\mu} - \bar{\mu}_F$.

If f_0 be the focal distance of a compound lens for line F, f_0', f_0'', &c. of the component lenses, then

$$\frac{1}{f} = \Sigma \frac{1}{f_0^{(r)}} + \Sigma \frac{b^{(r)}}{f_0^{(r)}} \cdot x + \Sigma \frac{b^{(r)} c^{(r)}}{f_0^{(r)}} \cdot x^2,$$

f being the focal length for the ray denoted by x.

If there be two lenses in the combination

$$\frac{1}{f} = \frac{1}{f_0} + \frac{1}{f_0} \frac{b' b''}{b'' - b'} (c' - c'') x^2.^*$$

Since the effect of changing the ray to be denoted by zero does not sensibly change the value of the coefficient $\dfrac{b' b''}{b'' - b'} (c' - c'')$, this may be taken as a measure of the irrationality of the combination.

Let there be three glasses (1), (2), (3); no combination free from secondary dispersion and of finite focal length can be made with these glasses if

$$\begin{vmatrix} 1, & 1, & 1 \\ b', & b'', & b''' \\ b'c', & b''c'', & b'''c''' \end{vmatrix} = 0.$$

Again, if the secondary chromatic aberration of (2) (1) is the same as that of (3) (1), then that of (2) (3) has also the same value, and the three glasses satisfy the above condition.

---

* That is, if the focal lengths be chosen for achromatism. [ED.]

Prof. Stokes has expressed the character of glasses in the following manner :—Let a prism of small angle i be perfectly achromatized by two prisms of standard glasses with angles i', i'' taken algebraically as regards sign, then

$$ai + a'i' + a''i'' = \text{deviation of any ray,}$$
$$abi + a'b'i' + a''b''i'' = 0,$$
$$abci + a'b'c'i' + a''b''c''i'' = 0 ;$$

hence

$$\frac{i'}{i''} = \frac{c'' - c}{c - c'} \frac{a''b''}{a'b'} .$$

If $c = c''$ this ratio is zero, but if $c = c'$ it is infinite; let $\frac{i'}{i''} = \tan \phi$, then the angle ϕ may be taken with a and b as a complete specification of the optical properties of the glass. Prof. Stokes's method has a great advantage in the close correspondence between the values of i, i', i'' and the powers of the component lenses of a perfectly achromatic object-glass, and also in the rapidity with which a determination can be made. The method adopted in this paper is convenient in the fact that a single standard glass is alone required.

The determinations were made with a spectrometer supplied to Messrs Chance Bros. & Co. by Mr Howard Grubb. The telescope and collimator are 2 inches aperture; the circle is 15 inches diameter, is graduated to 10', and reads by two verniers to 10″.

The lines of the spectrum observed were generally A, B, C, D, E, b, F, (G), G, h, H_1. D is the more refrangible of the pair of sodium lines, b is the most refrangible of the group of magnesium lines, (G) is the hydrogen line near G.

The method of smoothing the results by the aid of each other has been, first, to calculate a, b, and c from the mean values of μ for the lines B, F, H_1; second, to calculate values of μ from the formula obtained; third, to plot on paper the differences between μ observed and μ calculated, and to draw a free-hand curve among the points, and then inversely to take μ for each line from the curve.

It appeared desirable to express the standard values of $\bar{\mu}$, which are the means of those for Hard Crown, Soft Crown, Light

Flint, and Dense Flint, in terms of $\frac{1}{\lambda^2}$. In the following Table column

      I. gives λ, the wave-length in 10^{-4} centims.;

      II. the values of $\frac{1}{\lambda^2}$;

      III. the standard values of $\bar{\mu}$;

      IV. the values of $\bar{\mu}$ calculated from

$$\mu = a + b\,\frac{1}{\lambda^2} + c\,\frac{1}{\lambda^4},$$

where
$$a = 1\cdot539718,$$
$$b = 0\cdot0056349,$$
$$c = 0\cdot0001186;$$

      V. the differences of III. and IV.;

      VI. the values of $\bar{\mu}$ from the extended formula

$$\mu = a + b\,\frac{1}{\lambda^2} + c\,\frac{1}{\lambda^4} + d\,\frac{1}{\lambda^6},$$

where
$$a = 1\cdot538414,$$
$$b = 0\cdot0067669,$$
$$c = -\,0\cdot0001734,$$
$$d = 0\cdot000023;$$

      VII. the differences of III. and VI.

	I.	II.	III.	IV.	V.	VI.	VII.
B	·68668	2·12076	1·552201	1·552201	0·000000	1·552203	− 2
C	·65618	2·32249	1·553491	1·553444	+0·000047	1·553481	+10
D	·58890	2·88348	1·557030	1·556951	+0·000079	1·557033	− 3
E	·52690	3·60200	1·561612	1·561553	+0·000059	1·561613	− 1
b	·51667	3·74605	1·562530	1·562490	+0·000040	1·562538	− 8
F	·48606	4·23272	1·565692	1·565692	0·000000	1·565693	− 1
G	·43072	5·39026	1·573459	1·573536	−0·000077	1·573457	+ 2
h	·41012	5·94536	1·577356	1·577409	−0·000053	1·577349	+ 7
H_1	·39680	6·35121	1·580287	1·580287	0·000000	1·580289	− 2

It is interesting to remark that the curve representing μ in terms of $\frac{1}{\lambda^2}$ has a point of inflexion between C and D. An examination of the deviations from calculation for several glasses shows that probably all glasses exhibit a similar point of inflexion, the flints lower in the spectrum or in the ultra-red, and the crowns nearer to the middle of the visible spectrum. This fact may be of importance in the theory of dispersion when a detailed theory becomes possible; at least it is important as showing how unsafe it would be to calculate μ for very long waves or ultra-violet waves from any formula of three terms.

The following Tables of results need little or no further explanation; the first line gives the refractive indices finally obtained and regarded as most probable, the second line gives the values of μ from the formula of three terms

$$\mu - 1 = a \left\{ 1 + bx \left(1 + cx \right) \right\},$$

and the last gives the mean of the actual observations.

Hard Crown.—Two specimens of this glass were examined.

α. A prism from the three angles of which determinations were obtained, density 2·48575.

$$a = 0·523145. \quad b = 1·3077. \quad c = -2·33.$$

	A	B	C	D	E	b	F	(G)	G	h	H_1
Most probable value of μ	......	1·513625	1·514568	1·517114	1·520331	1·520967	1·523139	1·527994	1·528853	1·530902	1·532792
μ from formula............	......	1·513624	1·514560	1·517099	1·520327	1·520966	1·523145	1·528003	1·528362	1·530906	1·532789
Mean of observed values	1·511755	1·513624	1·514571	1·517116	1·520324	1·520962	1·523145	1·527996	1·528348	1·530904	1·532789

β. A prism with two angles of about 45°, density = 2·48664.

The first line gives the mean observed indices, and the second the differences from the most probable values of α.

	A	B	C	D	E	b	F	(G)	G	h	H_1
Mean of observed values	1·572296	1·514156	1·515105	1·517623	1·520860	1·521474	1·523657	1·528533	1·528856	1·531439	1·533318
	0·000541	0·000531	0·000537	0·000509	0·000529	0·000507	0·000518	0·000539	0·000503	0·000537	0·000526

Soft Crown.—The prism has three angles from which the mean is taken, density = 2·55035.

$$a = 0·5209904. \quad b = 1·4034. \quad c = -1·58.$$

	A	B	C	D	E	b	F	(G)	G	h	H₁
Most probable value of μ	......	1·510916	1·511904	1·514591	1·518010	1·518686	1·520996	1·526207	1·526595	1·529359	1·531416
μ from formula...........	......	1·510918	1·511900	1·514574	1·517991	1·518670	1·520994	1·526215	1·526603	1·529364	1·531418
Mean of observed values	1·508956	1·510918	1·511910	1·514580	1·518017	1·518678	1·520994	1·526208	1·526592	1·529360	1·531415

Titano-silicic Crown.—This glass was made on the suggestion of Professor Stokes in the hope of obtaining a glass of good quality which should be perfectly achromatic with a flint. The determinations were made on two angles; density = 2·55255.

$$a = 0·550466. \quad b = 1·5044. \quad c = -0·93.$$

	B	C	D	E	b	F	(G)	G	h	H₁
Most probable value of μ	1·539155	1·540255	1·543249	1·547088	1·547852	1·550471	1·556386	1·556830	1·559999	1·562392
μ from formula...........	1·539152	1·540249	1·543236	1·547074	1·547839	1·550466	1·556407	1·556852	1·560015	1·562388
Mean of observed values	1·539157	1·540253	1·543248	1·547088	1·547852	1·550471	1·556394	1·556825	1·559998	1·562392

Extra Light Flint Glass.—Determinations made from three angles, density = 2·86636.

$a = 0·549123. \quad b = 1·7064. \quad c = − 0·198.$

	A	B	C	D	E	b	F	(G)	G	h	H_1
Most probable value of μ	......	1·536450	1·537673	1·541011	1·545306	1·546166	1·549121	1·555863	1·556372	1·560010	1·562760
μ from formula............	......	1·536448	1·537663	1·540993	1·545296	1·546158	1·549123	1·555881	1·556390	1·560026	1·562760
Mean of observed values	1·534067	1·536450	1·537682	1·541022	1·545295	1·546169	1·549125	1·555870	1·556375	1·559992	1·562760

Light Flint Glass.—Determinations made from three angles, density = 3·20609.

$a = 0·583887. \quad b = 1·9605. \quad c = + 0·53.$

	B	C	D	E	b	F	(G)	G	h	H_1
Most probable value of μ	1·568558	1·570011	1·574015	1·579223	1·580271	1·583886	1·592190	1·592824	1·597332	1·600727
μ from formula............	1·568553	1·570009	1·574017	1·579226	1·580274	1·583887	1·592185	1·592815	1·597322	1·600724
Mean of observed values	1·568558	1·570007	1·574013	1·579227	1·580273	1·583881	1·592184	1·592825	1·597332	1·600717

Dense Flint.—Determinations from three angles, density = 3·65865.

$a = 0\cdot634744. \quad b = 2\cdot2694. \quad c = 1\cdot48.$

	B	C	D	E	b	F	(G)	G	h	H_1
Most probable value of μ	1·615701	1·617484	1·622414	1·628895	1·630204	1·634748	1·645267	1·646068	1·651840	1·656219
μ from formula	1·615700	1·617488	1·622427	1·628904	1·630210	1·634744	1·645258	1·646060	1·651837	1·656222
Mean of observations	1·615704	1·617477	1·622411	1·628882	1·630208	1·634748	1·645268	1·646071	1·651830	1·656229

Extra Dense Flint.—Determinations from only one angle, density = 3·88947.

$a = 0\cdot664226. \quad b = 2\cdot4446. \quad c = 1\cdot87.$

	A	B	C	D	E	b	F	(G)	G	h	H_1
Most probable value of μ and μ from formula	......	1·642874	1·644866	1·650388	1·657653	1·659122	1·664226	1·676111	1·677019	1·683577	1·688569
Observed	1·639143	1·642894	1·644871	1·650374	1·657631	1·659108	1·664246	1·676090	1·677020	1·683575	1·688590

Double Extra Dense Flint.—Determinations from two angles, density = 4·42162.

$a = 0\text{·}727237.$ $b = 2\text{·}7690.$ $c = 2\text{·}70.$

	A	B	C	D	E	b	F	(G)	G	h	H₁
Most probable value of μ and μ from formula	......	1·701060	1·703478	1·710201	1·719114	1·720924	1·727237	1·742063	1·743204	1·751464	1·757785
Mean of observations ...	1·696531	1·701080	1·703485	1·710224	1·719081	1·720908	1·727257	1·742058	1·743210	1·751485	

The following Table gives the value of $\dfrac{bb'}{b'-b}(c-c')$ for each glass when combined with the standard. It serves to show how little there is to choose between the glasses ordinarily used.

Hard Crown	Soft Crown	Titanic Crown	Extra Light Flint	Light Flint	Dense Flint	Extra Dense Flint	Double Extra Dense Flint
− 11·7	− 10·7	− 9·4	− 9·9	− 9·4	− 11·8	− 11·9	− 13·2

21.

ELECTROSTATIC CAPACITY OF GLASS.

[From the *Philosophical Transactions of the Royal Society*, Part I. 1878, pp. 17—23.]

Received May 17,—*Read June* 14, 1877.

1. IN his work on Electricity and Magnetism Professor Maxwell develops a theory in which electric and magnetic phenomena are explained by changes of position of the medium, the wave motion of which constitutes Light. He deduces with the aid of this theory that that velocity, which is the ratio of the electrostatic and electromagnetic units of electric quantity, is identical with the velocity of light. This deduction may be said to be verified within the limits of error of our knowledge of these quantities. He further finds that the product of the electrostatic capacity and the magnetic permeability of a transparent substance is equal to the square of the refractive index for long waves. The only available experiments for testing this result when Professor Maxwell's book was published[*] were the "Determinations of Electrostatic Capacity of Solid Paraffin," by Messrs Gibson and Barclay (*Phil. Trans.* 1871), and the "Determinations of Refractive Indices of Melted Paraffin," by Dr Gladstone. Considering

---

[*] Since then determinations have been made by Boltzmann for paraffin, colophonium, and sulphur (*Pogg. Ann.* 1874, vol. CLI. pp. 482 and 531, and vol. CLIII. p. 525), and for various gases (*Pogg. Ann.* 1875, vol. CLV. p. 403), by Silow for oil of turpentine and petroleum (*Pogg. Ann.* 1875, vol. CLVI. p. 389, and 1876, vol. CLVIII. p. 306), and by Schiller (*Pogg. Ann.* 1874, vol. CLII. p. 535) and Wüllner (*Pogg. Ann.* 1877, new series, vol. I. pp. 247, 361) for plate glass.

the difference in physical state in the two experiments the result verifies the theory fairly well. The various kinds of optical flint glass are suitable for the purpose of making a comparison of refractive indices and specific inductive capacity, since each is an article pretty constant in its composition and physical properties, and has small conductivity and return charge.

2. The only convenient form in which glass can be examined is a plate with plane parallel sides; this plate must form the dielectric of a guard ring condenser. Four instruments are thus required, the guard ring condenser, an adjustable condenser which can be made equal to the first, a battery for giving equal and opposite charges to the two condensers, and an electroscope to show when the added charges of the condensers are nil.

Guard Ring Condenser.—Fig. 1 represents the guard ring

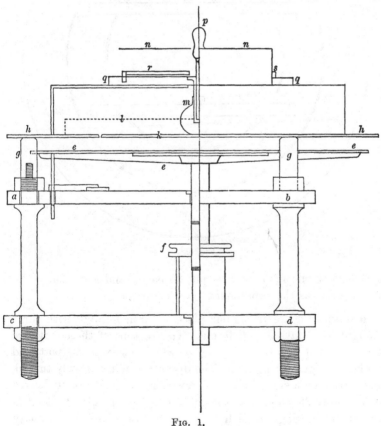

Fig. 1.

condenser in elevation; Fig. 2 in plan through *h h*. It consists essentially of an insulated brass disc *k* surrounded by a flat ring *h h*, and covered by a brass shield connected with *h h*. It is opposed by a larger disc *e e* parallel with *k* and *h h*, which is always connected to the case of the electrometer. The disc *k* and ring *h h* are connected, simultaneously charged, next separated,

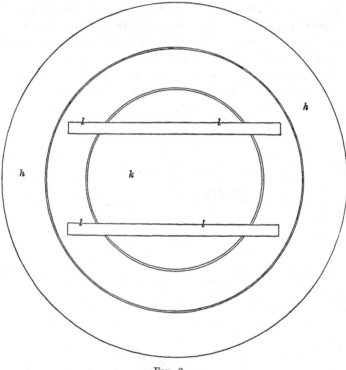

FIG. 2.

and then at one moment *h* is put to earth, and *k* discharged in such manner as the experiment may require.

a b and *c d* are triangular pieces of iron forming with three wrought-iron stays a stiff frame. To the tops of these stays are screwed three legs of ebonite *g g*, which serve to support and insulate the guard ring *h h*. The disc *e e* is of brass truly turned, it is carried on a stem which is screwed for a portion of its length with exactly 25 threads to the inch, a motion parallel to itself is secured by bearings in each frame plate; these are not ordinary round bearings which may work loose, but are of the form represented full size in Fig. 3. *e e* is prevented from rotating by a pin working through a hole in the upper triangular plate and pressed against one side of the hole by a steel spring. The plate *e e* is raised or lowered by a milled nut *f*, divided on the circumference into 100 parts, and bearing upon a piece of brass tubing secured to the lower plate of the frame. *k* is carried by two rods of ebonite *l l*, which insulate it from *h h*; both were faced in the lathe together so as to be truly in one plane. The diameter of the disc *k* is 150 millims., it is separated from the ring by a space of 1 millim. When the capacity of a glass plate is to be measured a dish of pumice and sulphuric acid is placed upon the disc *k* between the rods *l l*, and a second dish upon the triangular plate *c d*, the whole instrument being loosely surrounded by a glass

FIG. 3.

cylinder. This instrument also serves to measure with sufficient accuracy the thickness of the glass plates. To ascertain when the plates are in contact, or when the glass plate to be measured is in contact with *h k h*, slips of tissue paper are interposed between the ebonite legs *g g* and the plate *h h*, and the contact is judged by these slips becoming loose, a reading being taken for each slip.

The sliding condenser was the identical instrument used by Gibson and Barclay, kindly lent to the author by Sir W. Thomson; it was used simply as a variable condenser. Although a more finely graduated instrument than the guard ring condenser, it was not used as a measuring instrument, because its zero readings had to be valued by the guard ring condenser; it seemed better to use it like the counterpoise in the system of double weighing, adjusting it to the guard ring condenser with the glass in, then removing the glass and adjusting the guard ring condenser to equality with the sliding condenser. It suffices to say that the sliding condenser has two adjustments, a fine one denoted here by S_1, and a coarse one denoted by S_2.

The electroscope was Sir W. Thomson's quadrant electrometer adjusted for maximum sensibility and charged as highly as it would stand. A single Daniell's element gave from 120 to 160 divisions of the scale.

The battery consisted of 48 or of 72 Daniell's elements of a very simple construction; a piece of copper wire covered with gutta-percha is stripped for a short distance at each end, it is set in a test tube 6 or 7 inches long, a piece of zinc being soldered to its upper extremity. Some sulphate of copper in powder is put in the tube around the exposed wire, this is covered by a thick plug of plaster of Paris, and the element completed by the addition of dilute zinc sulphate solution, into which the zinc which is soldered to the wire of the next element dips. The element has a very high resistance, but that is of no consequence for electro-static experiments. The middle of the series is put to earth. The battery thus gives the means of charging two condensers to equal but opposite potentials. The poles of the battery are connected with the switch through the electrometer reversing key. In each case two experiments are made, one in which the guard ring is positive, in the other negative.

The switch is represented in plan in Fig. 4, and its place is indicated in elevation in Fig. 1. Calling the poles of the battery A and B, its purpose is to make rapidly the following changes of connexion :—

(1) A, sliding condenser; B, guard ring, disc k; earth, quadrant of electrometer.

(2) A, B, guard ring, earth; disc k, sliding condenser.

(3)   To connect the disc k and the sliding condenser to the quadrant of the electrometer.

The combination (1) may exist for any time long or short, but (3) follows (2) within a fraction of a second, and the observation of the electroscope consists in deciding whether or not the image moves *at the instant* of combination (3), and, if it moves, in which direction. In (2) the poles of the battery are put to earth, in order that one may be sure that the parts of the switch with

which they are connected do not disturb the result by inductive
action on the parts connected with
the condensers.

$q\,q$ is a plate of ebonite screwed
to the shielding cover of the condenser, r is a steel spring connected to earth, s a similar steel
spring connected to one pole of
the battery.

$t\,v$ are segments of brass of
which the securing screws pass
through to the brass cover.

$w\,u$, similar segments insulated from the brass cover and guard
ring connected respectively to
the sliding condenser and the
electrometer.

p is an ebonite handle and
brass pin which turns in an insulated brass socket connected by a
spring m with the disc k; p carries
a piece of ebonite $x\,x$ which moves
the springs $r\,s$ from contact with
$t\,v$ to contact with $u\,w$, and also
a spring $y\,y$ which may connect
$t\,v$ with the disc k, or, when turned
into the position indicated, w with
the disc k, and instantly after
both with the electrometer. One
pole of the battery is always connected to the guard $h\,h$. The
switch is protected against inductive action from the hand of
the observer, or from electrification of the top of the ebonite
handle when touched with the
finger, by a copper shield $n\,n$
connected with the guard ring
through the cover.

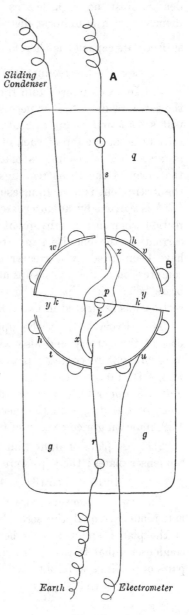

Fig. 4.

The guard ring screw reading is denoted by R. R (+) when the guard ring is positive, R (−) when it is negative. This condenser must be regarded as a circular plate of 151 millims. diameter with a uniform distribution of electricity on its under surface; its capacity is therefore $\dfrac{140\cdot3}{x}$ centimetres, where

$x = R$ − the reading when k and $e\,e$ are in contact.

3. In order to ascertain the distance between the plates from the screw reading R, it is necessary to know the reading when the plates $k\,h\,h$ and $e\,e$ are in contact. Slips of thin tissue paper are introduced at the top of each of the ebonite legs, the lower plate is raised, and a reading is made when each slip becomes loose; the mean of the three readings may be taken as the zero when the instrument is used to measure the thickness of plates, or when $h\,k\,h$ is carried by an interposed plate, but it will require a correction when in the subsequent measurements the upper plate is carried by the ebonite supports only, for the upper plate must have been lifted by a greater or less amount depending on the compression of the paper slips and on the imperfect rigidity of the brass before the slips can be released. The amount of this correction was estimated in two different ways.

1st. Everything on the upper plate was connected with one pole of the battery and also with the electrometer. The plates were brought to contact; it was found the slips became loose at 1·15, 1·13, 1·09, mean 1·12, the lower plate was very slowly lowered until the upper plate became insulated, as declared by the movement of the image on the electrometer scale. This occurred at 1·22, indicating a correction of 0·10.

2nd. A plate of light flint glass was introduced between the condenser plates; the slips were just loose at readings—

$$16\cdot15 \qquad 16\cdot16 \qquad 16\cdot11. \qquad \text{Mean } 16\cdot14.$$

The two condensers were now connected through the switch and rendered equal, the screw being turned to vary the distance of the plates, and the slide being adjusted to make the sliding condenser equal to the guard ring. The following corresponding pairs of readings were obtained:—

R	16·10	16·20	16·30	16·25	16·40
S_1	180	180	150	170	100
R	16·30	16·27	16·24	16·20	16·17
S_1	150	170	180	180	185

It thus appears that the capacity of the guard ring condenser does not begin to diminish till R is between 16·24 and 16·27. This indicates a correction between 0·10 and 0·13. Throughout the experiments a correction of 0·10 is used whenever the upper plate is carried by the ebonite legs alone.

4. The glasses examined were Chance's *optical light flint, double extra dense flint, dense flint, a special light flint,* and a piece of common plate glass.

Light flint, density about 3·2.

Two plates were examined of different thickness, the plates were also from different meltings of glass made at different times, and may be regarded as two quite independent samples of glasses intended to be of the same composition.

A. Thickness, 15·01 turns of the screw; diameter, 220 millims.

First Experiment.—Plates of guard ring condenser in contact with glass plate. 48 elements in the battery.

$$S_2 = 0 \qquad S_1 = 50 \text{ when sliding condenser positive.}$$
$$= 20 \qquad \text{,,} \qquad \text{,,} \qquad \text{negative.}$$
$$\text{Mean} = 35$$

It is found that $S_2 = 0$, $S_1 = 35$ is equal to the guard ring condenser with the glass plate out, when the distance between the plates is 2·18 turns of screw.

$$\text{Hence } K = 6·89.$$

Second Experiment.—Battery of 72 elements.

S_2 drawn out beyond the graduation,

$$S_1 = 160 \text{ when slide is positive.}$$
$$= 220 \qquad \text{,,} \qquad \text{negative.}$$
$$\text{Mean} = 190$$

Glass plate removed.

$$S_1 = 190 \qquad R(+) = 3·50 \qquad R(-) = 3·43.$$

Mean reading for contact of plates 1·14, when corrected 1·24.

So plate of glass 15·01 is equal to plate of air 2·225.

$$\text{Hence } K = 6·76.$$

B.   Thickness, 10·75 turns of screw; diameter, 220 millims.

Plates of guard ring both in contact with glass, battery of 72 elements.

$$S_2 = 25 \qquad S_1 = 450 \text{ when positive.}$$
$$= 400 \text{ when negative.}$$

Glass plate removed.

$$S_1 = 425 \text{ equivalent to } R (+) = 2·85$$
$$R (-) = 2·80$$

or plate of glass 10·75 equal to air 1·585.

$$K = 6·90.$$

Mean of three determinations—

6·85.

"*Double extra dense flint glass*," or "*Triple dense flint*," density about 4·5.

Thickness of plate, 24·27 turns; diameter, 235 millims.

First Experiment.—Plates in contact with glass.   48 elements in battery.

$$S_2 \text{ drawn out, } S_1 = 95.$$

Plate removed, condensers again equal when $R = 3·60$.

Hence $K = 10·28$.

Second Experiment.—Plates in contact with glass.   72 elements in battery.

$$S_2 \text{ drawn out, } S_1 = 55 \text{ when slide is positive.}$$
$$= 95 \qquad \text{,,} \qquad \text{negative.}$$
$$S_1 = 75 \text{ is equivalent to } R (-) = 3·61$$
$$R (+) = 3·69$$
$$K = 10·07.$$

The latter result is probably much the best; take 10·1 as most probable value.

In the next two glasses the determinations were made first with plates in contact with glass, second with a space of air between the glass and the upper plate; the results suggested the experiments of § 3.   In each case 72 elements were used.

Dense flint (the glass generally used in the objectives of tele-scopes).—Density about 3·66, thickness = 16·58 turns of the screw, diameter = 230 millims.

First Experiment.—Plates in contact with glass, S_2 drawn out.

When the slide is positive, $S_1 = 205$, on removal of glass plate this equals R (−) = 5·50. When the slide is negative, $S_1 = 175$, on removal of glass plate R (+) = 3·50.

$$\text{Hence } K = 7·34.$$

The mean zero reading being now 1·15.

Second Experiment.—R is put at 18·14 with glass between the plates.

S_2 drawn out when the slide is negative.

$S_1 = 10$ on removing glass equals R (+) 3·78, when the slide is positive.

$S_2 = 40$ on removing glass equals R (−) 3·79, *i.e.* glass 16·58 and air 0·32 are equivalent to air 2·525 or $K = 7·45$.

$$\text{Mean} = 7·4.$$

A very light flint.—Density about 2·87, thickness = 12·7 turns of the screw, diameter = 235 millims.

First Experiment.—Plates in contact with glass S_2 drawn out, when the slide is negative.

$S_1 = 380$ on removing glass equals R (+) 3·20, when the slide is positive.

$S_1 = 440$ on removing glass equals R (−) 3·18.

$$K = 6·6.$$

Second Experiment.—R was put at 14·50, S_2 was drawn out when the slide is positive.

$S_1 = 80$ on removing glass equal to R (−) 3·71, when the slide is negative.

$S_1 = 50$ on removing glass equal to R (+) 3·72, so glass 12·70 and air 0·55 is equivalent to air 2·475.

$$K = 6·55.$$

$$\text{Mean} = 6·57.$$

An attempt was made to determine K for a piece of plate glass; the considerable final conductivity of the glass caused no serious inconvenience, but the very great development of that polarization on which residual charge depends produced a condenser in which the capacity seemed to increase very rapidly indeed during a second or so after making connexions; this effect could not be entirely separated from the instantaneous capacity, a value $K = 7$ was obtained, but it was quite certain that a considerable part of this took time to develope.

5.   The repetition of the experiment in each case gives some notion of the probable error of the preceding experiments. Something must be added for the uncertainty of the contact reading. It will perhaps not be rash to assume the results to be true within 2 per cent.

Since the magnetic permeability cannot be supposed to be much less than unity, it follows that these experiments by no means verify the theoretical result obtained by Professor Maxwell, but it should not be inferred that his theory in its *more general* characters is disproved.

If the electrostatic capacities be divided by the density, we find the following quotients :—

	ρ	K	$\dfrac{K}{\rho}$	μ (index of refraction for line D)
Light flint...	3·2	6·85	2·14	1·574
Double extra dense ...	4·5	10·1	2·25	1·710
Dense flint	3·66	7·4	2·02	1·622
Very light flint... ...	2·87	6·57	2·29	1·541

Thus $\dfrac{K}{\rho}$ is not vastly different from a constant quantity. Messrs Gibson and Barclay find K for paraffin 1·977 ; taking the density of paraffin as 0·93, we have the quotient 2·13. This empirical result cannot of course be generally true, or the capacity of a substance of small density would be less than unity.

22.

ELECTROSTATIC CAPACITY OF GLASS AND OF LIQUIDS.

[From the *Philosophical Transactions of the Royal Society*, Part II. 1881, pp. 355—373.]

I. ELECTROSTATIC CAPACITY OF GLASS.

Received November 3,—Read December 16, 1880.

In 1877 * I had the honour of presenting to the Royal Society the results of some determinations of the specific inductive capacity of glass, the results being obtained with comparatively low electromotive forces and periods of charge and discharge of sensible duration.

In 1878 Mr Gordon† presented to the Royal Society results of experiments, some of them upon precisely similar glasses, by a quite different method, with much greater electromotive forces and with very short times of charge and discharge. Mr Gordon's results and my own are compared in the following table:—

	Gordon		Hopkinson, 1877
	Christmas, 1877	July and Aug., 1879	
Double extra-dense flint	3·164	3·838	10·1
Extra-dense flint	3·053	3·621	
Light flint	3·013	3·443	6·85
Hard crown	3·108	3·310	

* *Phil. Trans.* 1878, Part I.
† *Ib.* 1879, Part I.

It is quite clear that such enormous differences cannot be due to mere errors of observation; they must arise from a radical defect in one method or the other, or from some property of the material under investigation. I have now repeated my own experiments with greater battery power, and with a new key for effecting the connexions of the condensers, and have obtained substantially the same results as before.

Two hypotheses suggest themselves as to the physical properties of glasses which might, if true, account for the diversity of results :—(i) In my own earlier experiments a considerable time elapsed, during which some have thought residual charge might flow from the glass condenser and go to swell the capacity determined. Sir W. Thomson had informed me that experiments had proved that the capacity of a good insulating glass is sensibly the same, whether the period of discharge be the ten- or twenty-thousandth of a second, or say one-quarter of a second. This statement has been verified. (ii) It appeared plausible to suppose that specific inductive capacity of glass was not a constant, but was a function of the electromotive force—in other words, that the ratio $\dfrac{\text{charge of glass condenser}}{\text{difference of potential}}$ was less when the electromotive force was great than when it was small. This surmise gains some force from Dr Kerr's electro-optical results, which show that electrostatic and optical disturbance of a dielectric are not superposable. It has, however, been submitted to a direct test, with the result that, within the limits tried, specific inductive capacity *is* a constant, and that it is not possible that the discrepancy of experimental results can be thus explained. Finally, I have made a rough model of Mr Gordon's five-plate balance, and used it to make determinations of specific inductive capacity.

Firstly, a brass plate was tried, and its capacity was found less than unity instead of infinite.

Secondly, by varying the distances of the plates of the balance from each other, different values of the specific inductive capacity of the same glass were obtained. In fact, it has been shown that the five-plate induction balance cannot be freely relied upon to give correct values of specific inductive capacity.

I conclude that the values I published in 1877 are substantially accurate, whether the period of discharge be $\frac{1}{20000}$ or $\frac{1}{2}$ second, whether the electromotive force be one volt per millimetre or 500 volts per millimetre, and that Mr Gordon's different result is to be explained by a defect in the method he used.

(I.) *To prove that a condenser of well-insulated glass may be almost completely discharged in $\frac{1}{20000}$ second.*

For this experiment it is essential that the effect of conduction over the surface of the glass should be insensible. A jar, such as that used in Sir W. Thomson's electrometer, is unsuitable. The proper form for the condenser is a flask with a thin body and a thick neck, filled with strong sulphuric acid to the neck. Such a flask of light flint glass was prepared, and was instantaneously discharged in the following manner:—The interior of the flask was connected to a metal block, A. Upon this block rests a little L-shaped metal piece, B, which can turn on a knife-edge, C. A

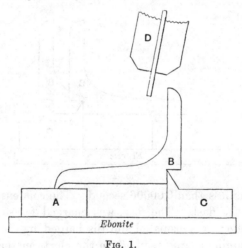

Fig. 1.

and C are carried on a block of ebonite, and are therefore insulated. D is a piece of metal connected to earth, and rigidly attached to the extremity of a pendulum. The pendulum is drawn aside and let go; the piece D strikes B and puts the jar to earth, and instantly afterwards breaks the contact with A, and drives away the piece B. In all cases the pendulum was drawn aside 45°, and in all the experiments but one mentioned below it

5—2

made 93 half-oscillations per minute. The duration of the discharge was determined by the following method, which I arranged for myself, unaware that a similar method had been used by Mr Sabine*. A condenser of known capacity is connected to *A* through a known resistance; the condenser receives a known charge whilst connected to the electrometer; the piece *B* is struck by the pendulum, and the remaining charge is observed. Two experiments were made; in each the condenser was of tinfoil and paraffin, such as are used by Messrs Clark, Muirhead, and Co. for telegraph purposes, and had a capacity of 0·29 microfarad. The resistances were respectively 512 ohms and 256 ohms. The results gave respectively duration of discharge 0·0000592 second and 0·0000595 second. We may take it that the duration of

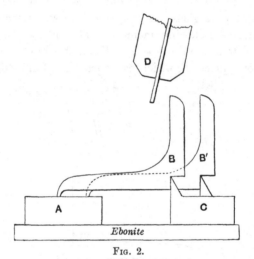

Fig. 2.

discharge was less than 0·00006 second. The condenser was now replaced by the flask. The flask was charged for some seconds from the battery, was insulated and discharged by the pendulum, and the remaining charge read off on the electrometer so soon as the image came to rest. In a first experiment the charge was from four elements (= 444 divisions of the scale), and the charge remaining gave deflection 34 divisions. In a second experiment the charge was from eight elements (= 888 divisions), and the remaining charge was 61 divisions. Even this small residual charge is largely due to the inductive action of the needle of the

* *Phil. Mag.*, May, 1876.

electrometer on the quadrant connected to the flask. To prove
this, the experiment was varied by beginning with the quadrant
separated from the flask, and only connecting these after discharge had been made. With eight battery elements, the remaining charge in the flask was found to be 25 divisions; with 20
elements, 61 divisions. From these experiments we may conclude that, if a flask of light flint glass be charged for some
seconds and be discharged for 0·00006 second, the residual charge
coming out in the next three or four seconds is certainly less than
3 per cent. of the original charge. It was important to learn if
this 3 per cent. was sensibly diminished if the time of discharge
was somewhat increased. For this purpose the time of oscillation was increased, and the arrangement of piece B and knife-
edge C was duplicated, so that the flask was twice discharged
within an interval of about $\frac{1}{80}$ second between. The result was,
with charge from eight elements and the flask initially connected
to the quadrant, a remaining charge of 61 divisions, exactly the
same as when the discharge only lasted $\frac{1}{17000}$ second. I conclude
that, with this glass, it matters not whether the discharge of the
flask last $\frac{1}{17000}$ second or $\frac{1}{80}$ second; its capacity is the same.
This is in precise accord with what Sir W. Thomson told me
before I began the experiments for my former paper.

(II.) *Determinations with the guard-ring condenser*.*

It has been suggested that my former results were liable to
uncertainty from the small potentials used and from the comparatively long time of discharge. The main purpose of the
present experiments has been to ascertain the force of the objec-
tions. As the principle of the method is the same as in the
earlier paper, it is only needful to explain the alterations the
apparatus has undergone.

The battery.—A chloride of silver battery of 1000 elements
was constructed and very carefully insulated, both as regards cell
from cell and tray from tray. Each tray contained 50 cells and
the set of 20 trays was conveniently enclosed in a wooden case
provided with suitable terminals. As my experience of the

---

* The cost of the additional instruments used has been defrayed by a Royal
Society Grant. The battery and some of the other instruments were made by
Messrs L. Clark, Muirhead, and Co., the remainder by Mr Groves.

battery is but short I shall not now minutely describe its details; it is sufficient to say that by connecting its middle to earth two condensers can be charged to equal and opposite potentials of 500 elements.

The guard-ring condenser.—This is the instrument of my former experiments, with the switch removed and some slight improvements in mechanical detail. It is by no means perfect in workmanship, and the irregularities of the results now to be given are to be attributed to such imperfections. It was not worth while to make a new instrument, as, for any present interest, determinations of capacities of glasses, correct to 1 per cent., are as valuable as if they were absolutely accurate.

The sliding condensers.—Two sliding condensers were constructed possessing together a very considerable range of capacity. Each has a single scale and is used as before merely as a balance to the guard-ring condenser, excepting in one experiment, the subject of the next section.

The switch.—The switch formerly used performed the following operations:—Initially, the quadrant of the electrometer was to earth, the guard-ring and the plates of the guard-ring condenser were connected to one pole of the battery, the sliding condenser to the other pole. On turning the handle the quadrant and the condensers were insulated; next, the charges of the condensers were mixed, the guard-ring being put to earth at the same time; and, finally, the connected condensers were connected to the quadrant of the electrometer; they remained so connected until the handle of the switch was turned back into its first position. This instrument could not be used to determine capacities when the residual charge was great, as in the case of plate glass, and was unsatisfactory to anyone who held that flint glass condensers discharged very much more in a time comparable with one second than in a minute fraction of a second. The new switch was arranged to effect the further operation of breaking contact between the condensers and the quadrant immediately after the contact was made. It is also arranged for much higher insulation, the old switch being quite useless for the greater battery power used.

The whole switch, binding screws and all, is covered with a brass cover connected to earth and provided with apertures for the connecting wires. The connecting wires are insulated with gutta-percha, covered with a metallic tape as an induction shield, this tape being of course connected to earth.

The mode of experiment was substantially as before. A glass plate was introduced in the guard-ring condenser, and the sliding condenser adjusted till the capacities were equal; the glass plate was removed and the guard-ring condenser, with air as its only dielectric, was adjusted till its capacity was equal to that of the sliding condenser. In every case the battery was reversed and the mean taken.

The following tables give the results obtained :—

All measures are given in terms of turns of the micrometer screw of the guard-ring condenser, of which there are 25 to the inch.

Column I. gives the circumstances of the particular experiment.

Column II. the distance between the plates of the condenser with glass in.

Column III. the same distance with air only when the capacity is the same as in II.

Column IV. the thickness of air plate equivalent to glass plate.

Column V. resulting value of K.

DOUBLE EXTRA-DENSE FLINT. Density, 4·5. Thickness of plate, 24·27.

I.	II.	III.	IV.	V.
200 elements used, 100 to each condenser, glass in contact with both plates	24·27	2·48	2·48	9·78
1000 elements, contact with both plates	24·27	2·48	2·48	9·78
1000 elements, resting on lower plate, space between glass and upper plate......................	24·69	2·865	2·445	9·92
Ditto ditto ditto	25·19	3·36	2·44	9·94
Ditto ditto ditto	25·39	3·57	2·45	9·90
Glass separated from lower plate by three small pieces of ebonite also separated from upper plate	25·19	3·36	2·44	9·94

Mean of last five experiments, $K = 9·896$.
Result formerly published, 10·1.

DENSE FLINT. Density, 3·66. Thickness of plate, 16·57.

I.	II.	III.	IV.	V.
Glass in contact with both plates, 400 elements	16·57	2·265	2·265	7·31
Glass in contact with both plates, 1000 elements	16·57	2·265	2·265	7·31
Glass resting on lower plate, 1000 elements	17·19	2·85	2·23	7·43
Ditto ditto	17·69	3·36	2·24	7·39

Mean of last three experiments, $K = 7·376$.

Result formerly published, 7·4.

LIGHT FLINT. Density, 3·2. Thickness of plate, 15·04.

I.	II.	III.	IV.	V.
Glass in contact with both plates, 1000 elements	15·04	2·215	2·215	6·79
Glass resting on lower plate, 1000 elements	15·29	2·505	2·255	6·67
Ditto ditto	15·69	2·865	2·215	6·79
Ditto ditto	16·19	3·42	2·27	6·62

Mean value of $K = 6·72$.

Results formerly published, 6·89 and 6·76 = 6·83.

LIGHT FLINT. Thickness, 10·75.

I.	II.	III.	IV.	V.
Contact with both plates, 1000 elements	10·75	1·61	1·61	6·67
Resting on lower plate	11·19	2·035	1·595	6·74
Ditto	11·69	2·555	1·615	6·65

Mean value of $K = 6·69$.

Result formerly published, 6·90.

Mean result for light flint, 6·72.

Mean formerly published, 6·85.

VERY LIGHT FLINT. Density, 2·87. Thickness, 12·70.

I.	II.	III.	IV.	V.
Glass in contact with both plates, 400 elements	12·7	1·915	1·915	6·63
Glass in contact with both plates, 1000 elements	12·7	1·915	1·915	6·63
Glass in contact with lower plate only, 1000 elements	12·99	2·215	1·925	6·59
Ditto ditto ditto	13·39	2·61	1·920	6·61

Mean of last three, $K = 6·61$.
Result formerly published, 6·57.

HARD CROWN. Density, 2·485. Thickness, 11·62.

I.	II.	III.	IV.	V.
Glass in contact with both plates, 1000 elements	11·62	1·675	1·675	6·93
Glass in contact with lower plate only, 1000 elements	11·70	1·74	1·66	7·0
Ditto ditto ditto	11·90	1·945	1·665	6·98
Ditto ditto ditto	12·30	2·255	1·675	6·93

Mean value of $K = 6·96$.

PLATE GLASS. Thickness, 6·52.

I.	II.	III.	IV.	V.
Glass in contact with lower plate only, 400 elements	7·70	1·95	0·77	8·47
Glass in contact with lower plate only, 1000 elements	7·70	1·95	0·77	8·47
Ditto ditto ditto	7·40	1·665	0·785	8·43

Mean value of $K = 8·45$.

REMARK.—On account of the small thickness of the equivalent plate of air, $\frac{1}{30}$ inch, this result is subject to a greater probable error than the others. No inconvenience or uncertainty was experienced from the effect of residual charge. If the switch be arranged so that contact with the electrometer is not broken, observation becomes at once impossible.

These results show that my former experiments require no material correction, except in the case of plate glass, for which an accurate experiment was formerly impossible. They also show that electrostatic capacity does not depend on electromotive force up to 200 volts per centimetre for double extra-dense flint, and a somewhat higher electromotive force for the other glasses. It is desirable to show that the same is true for a wider range.

PARAFFIN.   Thickness, 20·19.

I.	II.	III.	IV.
Resting on lower ...	23·82	12·42	8·79
Ditto ...	22·71	11·32	8·80
Ditto ...	21·37	9·96	8·78
Contact with both...	20·19	8·78	8·78

Mean value of $K = 2·29$.

In this case the guard-ring condenser was always charged with 700 elements, the slide with 300 in order that the same sliding condenser might be used.

Boltzmann gives 2·32 for paraffin for short times of discharge.

(III.)   *To show that K is a constant, as is generally assumed.*

Dr De La Rue very kindly allowed me to try a few preliminary experiments last February with his great chloride of silver battery. A flask of extra-dense flint glass was used, insulated with sulphuric acid precisely as in my experiments on residual charge. The comparison was made with a large sliding condenser having a scale graduated in millimetres. Taking one division of the scale (= about 0·0000026 microfarad) as a temporary unit of capacity, I found it impossible to say whether the capacity of the flask was greater or less than 390 divisions, whether the charge in each condenser was 20 elements or 1800 elements. Subsequently a similar experiment was tried with my own battery and a flask of light flint, with the following results, each being the mean of four readings :—

Charge to each condenser in AgCl elements.	Capacity in millim. divisions of sliding condenser.
10	273·75
100	274·00
200	273·75
300	274·5
400	273·0
500	273·5

The mean of these is 273·75, and the greatest variation from the mean 0·28 per cent.

The conclusion has some considerable importance, for some conceivable molecular theories of specific inductive capacities would lead to the result that capacity would be less when the charge became very great, as is actually the case with the magnetic permeability of iron (*vide* Maxwell, vol. II. chap. 6).

The flasks tried are about 1 millim. thick.   I intend to try a very thin glass bulb, testing it to destruction.

[In order to extend the limits of this test, a thin bulb 29 millims. diameter was blown on a piece of thermometer tube and its capacity compared with the sliding condenser with varying charge, as follows:—

100 battery elements to each condenser, capacity of bulb was 297 scale-divisions.

300 elements, capacity = 297 divisions.

500 elements, capacity = 297½ divisions.

The bulb was afterwards broken and the thickness of the fragments measured; they ranged from 0·05 to 0·15 millim., the major portion being about 0·1 millim.   We may conclude with confidence that the value of K for the glass tested continues constant up to 5000 volts per millimetre.—Dec. 9, 1880.]

An experiment was subsequently tried to ascertain if specific inductive capacity varied with the temperature of the dielectric. Accurate results could not be obtained, owing to the expansion of the acid, causing it to rise in the neck of the flask.   The result of the single experiment tried was, however, that the flask at 14° C. had a capacity equal to 275 divisions of the sliding condenser; at 60° C. it was equal to 280 divisions.   Having regard to the increase of capacity due both to the expansion of the glass (which may safely be neglected) and to the expansion of the acid (which is material), we can only conclude that the capacity of glasses certainly does not change rapidly with temperature—that consideration of temperature cannot be expected to reconcile Professor Maxwell's theory with the results of experiment.

I have repeated the temperature experiment with greater care. The flask was cleansed, filled a little short of the shoulder with acid, and arranged for heating and testing as before. In order to avoid the effect of rising of the level of the acid from expansion, the flask was heated to its highest temperature before any observation. It was assumed that on cooling the surface of the flask would continue to conduct to the level at which the acid had been.

The following table gives the results of the experiment:—

Temperature Centigrade.	Capacity.	
81	$269\frac{1}{2}$	11th Nov.
48	266	,,
27	$263\frac{1}{2}$	,,
12	262	12th Nov.
$39\frac{1}{2}$	$266\frac{1}{2}$	,,
$67\frac{1}{2}$	$268\frac{1}{2}$	,,
83	$271\frac{1}{2}$	,,
60	268	,,
$50\frac{1}{2}$	267	,,
13	264	13th Nov.

We may conclude, I think safely, that the specific inductive capacity of light flint does increase slightly, but that the increase from 12° to 83° does not exceed $2\frac{1}{2}$ per cent. The conductivity of the same glass* increases about 100-fold between the same temperatures, and the residual charge also increases greatly.

(IV.) *Examination of the method of the five-plate induction balance.*

The theoretical accuracy of this method rests on the assumption that the distance between the plates may be considered small in comparison with their diameter. When this condition is not

---

* " Residual Charge of the Leyden Jar," *Phil. Trans.* 1877; *supra*, p. 31.

sufficiently considered, it is easy to see that it is not likely that correct results will in all cases be obtained; for suppose that in lieu of the plate of glass a thin sheet of metal of considerable size is interposed between the fourth and fifth plates of the balance, it ought to be needful to withdraw the fifth plate by an amount equal to the thickness of the sheet. One can apprehend that it will be actually necessary to push it in, but to an extent which it would not be easy to calculate.

Some doubt is also thrown upon the practical accuracy of the method by the fact that Mr Gordon has arrived at the very unexpected result that the specific inductive capacities of glasses change with the lapse of time.

In order to satisfy myself on the point I had a rough model of a five-plate induction balance made. The instrument is far too rough to give minutely accurate results if the method were good, but I believe it is sufficient to show rapidly that it cannot be used with safety. The insulation was not perfect, and no attempt was made to enclose the instrument or shield the connexions from casual inductive action. The plates are all 4 millims. thick; they are, as in Mr Gordon's apparatus, 6 and 4 inches diameter. Each plate is suspended in a vertical plane by two rods and hooks from two of a set of four horizontal rods of varnished glass. The plates can thus be placed parallel to each other at any distance apart that may be desired. The distance between the plates was measured by a pair of common callipers and a millimetre rule to the nearest $\frac{1}{4}$ millimetre. For convenience, let the plates be named A, B, C, D, E, as in the accompanying diagram. In a first experiment B and D were respectively connected to the quadrants of an electrometer of which the jar was charged in the usual way. A and E were connected to one pole of a battery of 20 AgCl elements, C to the other pole through an ordinary electrometer reversing key, E was adjusted till the disturbance of the image was a minimum, when the key was reversed. This method was unsatisfactory, probably because in the act of reversing all the plates A, C, E were momentarily connected to one of the poles, and also because the insulation of the plates B, D was imperfect. The experiments, however, sufficed to prove beyond doubt that the instrument gave diminishing values to the specific inductive capacity of glass, as the distance of the five plates from each other was increased from 12 millims. to 32 millims., also that it gave values less than unity for the specific inductive capacity of brass in the form of a plate 3·5 millims. thick. More satisfactory working was attained by approximating, so far as my instruments admitted, to the methods of Mr Gordon. *B* and *D* were, as before, connected to the quadrants, *C* was connected to the interior of the jar and to one pole of an ordinary induction coil; *A* and *E* to the case of the instrument and to the other pole of the induction coil. The plate *E* was adjusted till the working of the coil caused no deflection of the image on the scale. In each case the plate examined was placed approximately half-way between *D* and *E*.

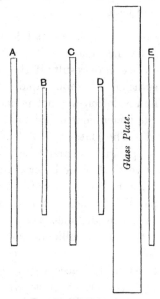

Fig. 3.   Half full size.

The following table gives results of a plate of double extra-dense flint 24·75 millims. thick and 235 millims. diameter, and of a plate of brass 3·5 millims. thick and 242 millims. diameter.

Column I. gives the air-space between the plates *AB*, *BC*, or *CD*.

Column II. the air-space *DE* (a_1) when no dielectric plate was present.

Column III. the distance *DE* (a_2) when a dielectric was introduced.

Column IV. the value of the difference $b - (a_2 - a_1)$, b being the thickness of the plate, which ought to be constant for each plate.

Column V. the specific inductive capacity $= \dfrac{b}{b - (a_2 - a_1)}$.

DOUBLE EXTRA DENSE FLINT, 24·75 mm. thick.

I.	II.	III.	IV.	V.
5	$5\frac{1}{4}$	27	3	8·25
8	$8\frac{1}{4}$	$30\frac{1}{4}$	$2\frac{3}{4}$	9·0
12	$11\frac{3}{4}$	$31\frac{3}{4}$	$4\frac{3}{4}$	5·21
18	21	$37\frac{1}{4}$	$8\frac{1}{2}$	2·91
25	$32\frac{1}{2}$	$43\frac{3}{4}$	$13\frac{1}{2}$	1·83
32	$44\frac{1}{2}$	$49\frac{1}{2}$	$19\frac{3}{4}$	1·25

True value of $K = 9·896$.

BRASS PLATE, 3·5 mm. thick.

I.	II.	III.	IV.	V.
5	4·5	6·75	1·25	2·8
8	8·0	6·25	5·25	0·66
12	11·25	10·0	4·75	0·73
32	44·5	16·5	31·5	0·11

True value of $K = \infty^{\text{ty}}$.

Inspection of the column IV. shows how impossible it is to attribute the variations of K to any mere error of observation even with the roughest appliances. Column V. demands no comment.

II. ELECTROSTATIC CAPACITY OF LIQUIDS*.

Received January 6,—Read January 27, 1881.

The number of substances suitable for an exact test of Professor Maxwell's electromagnetic theory of light is comparatively limited. Amongst solids, besides glass, Iceland spar, fluor-spar,

* The abstract of this paper is published in the *Proceedings* under the title " Dielectric Capacity of Liquids."

and quartz have been examined by Romich and Nowak[*], giving results for specific inductive capacity much in excess of the square of the refractive index. On the other hand, the same observers, with Boltzmann, obtain for sulphur a value of the capacity in reasonable accord with theory.

On liquids the only satisfactory experiments are those of Silow[†] on turpentine and petroleum oil, in which the capacity is precisely equal to the square of the refractive index for long waves.

Silow finally obtains for long waves and capacity—

	$\mu_\infty.$	$\sqrt{K}.$
Turpentine	1·461	1·468
Petroleum I.	1·422	1·439
Petroleum II.	1·431	1·428
Benzol	1·482	1·483

A comparison of the whole of the substances which have been examined indicates the generalisation that bodies similar in chemical composition to salts, compounds of an acid, or acids and bases, have capacities much greater than the square of the refractive index, whilst hydrocarbons, such as paraffin and turpentine, cannot be said with certainty to differ from theory one way or the other. It seemed desirable to test this conclusion by experiments on animal and vegetable oils and on other paraffins. It was probable that the compounds of fatty acids and glycerine would have high capacities.

Samples were tested of colza oil, linseed oil, neatsfoot oil, sperm oil, olive oil, castor oil, turpentine, bisulphide of carbon, caoutchoucine, the paraffin actually in use for the electrometer lamp, and three widely different mineral oils kindly given to me by Mr F. Field, F.R.S., to whom I am indebted for the boiling points given below.

The method of experiment was very simple. The sample was first roughly tested for insulation. It was found that it was useless to attempt the samples of colza or linseed oils, of caoutchoucine, or of bisulphide of carbon, but that the rest had sufficient insulation for the tolerably rapid method I was able to use.

[*] *Wiener Sitzb.* vol. LXX. Part II. p. 380.
[†] *Pogg. Ann.* vol. CLVI. 1875, p. 389, and CLVIII. 1876, p. 313.

The fluid condenser consisted of a double cylinder to contain the fluid, in which an insulated cylinder could hang; three brass rods suspended the latter from an ebonite ring which rested on three legs rising from the outer cylinder of the annular vessel. The position of the insulated cylinder was geometrically determined by three brass stops (*a, a, a*) which abutted against the legs which carried the ring, six points being thus fixed. A dummy ebonite ring with three brass rods, but without the cylinder, was provided for the purpose of determining the capacity of all parts and connexions not immersed in fluid.

The condenser was balanced against a sliding condenser, first with air and then with fluid.

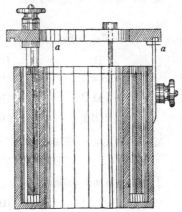

FIG. 4. Half full size.

The key which was used for experiments on plates was used here also, leaving the piece connected to the guard-ring idle.

The capacity of the sliding condenser was first tested with the result that to the reading of the slide 82·2 must be added to obtain the capacity in terms of the millimetre divisions of the scale. The capacity of the fluid condenser empty, with its connexions, was 106·5 divisions. The capacity of the dummy and connexions was 7·7, so that the nett capacity of the fluid condenser was 98·8. In all cases 1000 AgCl elements were tried, these being divided between the two condensers.

The following tables give the results obtained :—

Column I. is the number of elements charging the fluid condenser, the complement being used on the sliding condenser.

Column II. the reading of the slide plus 82·2 when a balance was obtained; this is the mean of two readings when the fluid condenser was respectively charged positive and negative.

Column III. is the capacity calculated from the experiment.

Petroleum spirit.   Boiling point, 159°.

I.	II.	III.
400	133·2	1·94
500	196·7	1·91
600	294·7	1·91

Mean value of $K = 1·92$.

Petroleum oil (Field's).   Boiling point, 310°.

I.	II.	III.
350	114·2	2·07
400	141·2	2·06
500	212·2	2·07

Mean value of $K = 2·07$.

Petroleum oil (common).

I.	II.	III.
400	144·2	2·11
500	214·2	2·09
600	321·2	2·09

Mean value of $K = 2·10$.

Ozokerit lubricating oil.   Boiling point, 430°.

Two determinations of this oil were made some days apart; at the time of the first determination the oil was slightly turbid. In the interval before determining the refractive index the upper portion became clear, the heavier particles having settled down. The capacity of the clear oil was then determined, and the results are given in the second table. It is possible that if the oil remain quiescent for a longer time a further reduction may be observed.

First experiment.

I.	II.	III.
400	149·2	2·19
500	223·2	2·18
600	334·7	2·18

Mean value of $K = 2·18$.

Second experiment.

I.	II.	III.
400	146·2	2·14
500	217·7	2·12
600	327·7	2·13

Mean value of $K = 2·13$.

Olive oil.

I.	II.	III.
300	137·7	3·17
400	213·7	3·16
500	319·2	3·15

Mean value of $K = 3·16$.

Castor oil.

I.	II.	III.
250	160·2	4·78
300	306·2	4·79
500	478·7	4·76

Mean value of $K = 4·78$.

Sperm oil.

I.	II.	III.
300	132·2	3·04
400	202·7	3·00
500	306·7	3·02

Mean value of $K = 3·02$.

Neatsfoot oil.

I.	II.	III.
300	134·2	3·09
400	206·7	3·06
500	311·2	3·07

Mean value of $K = 3·07$.

Turpentine.

A satisfactory determination for turpentine was not obtained. The turpentine seemed to act on the material of the vessel. After being in the condenser a short time its insulation was much reduced. When the charge had a potential of about 600 elements the condenser discharged itself disruptively through the turpentine. However, with a charge of 100 elements on each condenser a balance was obtained at 228·2, indicating a specific inductive capacity 2·23.

The refractive indices were determined *from the same samples* as the capacities in the usual way by the minimum deviation of a fluid prism. The spectrometer was the same I had previously used for experiments on glass (*Proc. Roy. Soc.* 1877). The observations were made for the hydrogen lines and the sodium lines, from these the index for long waves was calculated by the formula $a + \dfrac{b}{\lambda^2}$. The results are given in the following table :—

	μC	μD	μF	μG	$\mu \infty$	Temperature
Petroleum spirit............	1·3952	1·3974	1·4024	1·4065	1·3865	12·75
Petroleum oil (Field's) ...	1·4520	1·4547	1·4614	1·4670	1·4406	13·0
Petroleum oil (common)	1·4525	1·4551	1·4615	1·4670	1·4416	13·0
Ozokerit lubricating oil .	1·4558	1·4585	1·4653	. .	1·4443	13·0
Turpentine	1·4709	1·4738	1·4811	1·4871	1·4586	13·25
Castor oil....................	1·4785	1·4811	1·4877	1·4931	1·4674	13·5
Sperm oil....................	1·4724	1·4749	1·4818	. .	1·4611	13·75
Olive oil	1·4710	1·4737	1·4803	. .	1·4598	14·0
Neatsfoot oil	1·4673	1·4696	. .	. .	1·4578	14·0

In the following table is given a synoptic view of the comparison of $\mu\infty^2$ and K:—

	$\mu\infty$	$\mu\infty^2$	K
Petroleum spirit	1·3865	1·922	1·92
Petroleum oil (Field's)......	1·4406	2·075	2·07
Petroleum oil (common) ...	1·4416	2·078	2·10
Ozokerit lubricating oil ...	1·4443	2·086	2·13
Turpentine....................	1·4586	2·128	2·23
Castor oil	1·4674	2·153	4·78
Sperm oil	1·4611	2·135	3·02
Olive oil..................:.....	1·4598	2·131	3·16
Neatsfoot oil	1·4578	2·125	3·07

A glance shows that while vegetable and animal oils do not agree with Maxwell's theory, the hydrocarbon oils do. It must, however, never be forgotten that the time of disturbance in the actual optical experiment is many thousands of million times as short as in the fastest electrical experiment even when the condenser is charged or discharged for only the $\frac{1}{20000}$ second.

23.

ON THE REFRACTIVE INDEX AND SPECIFIC INDUCTIVE CAPACITY OF TRANSPARENT INSULATING MEDIA *.

[From the *Philosophical Magazine*, April, 1882, pp. 242—244.]

ONE of the deductions from Maxwell's electromagnetic theory of light is, that the specific inductive capacity of a medium is equal to the square of its refractive index. Another deduction is, that a body which is opaque to light, or, more generally, to radiant energy, should be a conductor of electricity. The first deduction appeared so clear an issue that many experimenters have put it to the test. The results may be briefly summarized thus:—Some bodies (such, for example, as hydrocarbon oils and † paraffin-wax) agree with Maxwell's law so well that the coincidence cannot be attributed to chance, but certainly points to an element of truth in the theory: on the other hand, some bodies, such as glass‡ of various kinds, fluor-spar§, Iceland spar§, and the animal and vegetable oils‖, have specific inductive capacities much greater than is indicated by their refractive indices.

How do these latter results really bear on Maxwell's theory? The facts are these. Taking the case of one substance as typical,

---

* Read before the Physical Society on February 25, 1882.

† Silow, *Pogg. Ann.* 1875, p. 382; 1876, p. 306. Hopkinson, *Phil. Trans.* 1881, Part II. p. 371; *supra* p. 74.

‡ "Cavendish Researches," edited by Clerk Maxwell; Schiller, *Pogg. Ann.* 1874, p. 535; Wüllner, *Sitz. k. bayer. Akad.* 1877, p. 1; Hopkinson, *Phil. Trans.* 1878, Part I., 1881, Part II.

§ Romich and Nowak, *Wiener Sitz.* Bd. LXX. Part II. p. 380.

‖ Hopkinson, *Phil. Trans.* 1881, Part II.; *supra* p. 82.

the refractive indices of light flint-glass are very accurately known, the period of disturbance ranging from $\dfrac{1}{4\cdot0 \times 10^{14}}$ second to $\dfrac{1}{7\cdot6 \times 10^{14}}$ second; the specific inductive capacity is known to be about 6·7, the time of electrical disturbance being from $\frac{1}{17000}$ second to a few seconds. If from the observed refractive indices we deduce by a formula of extrapolation the refractive index for very long waves, we find that its square is about one-third of 6·7. There can be no question about the accuracy of the observed refractive indices; and I have myself no doubt about the specific inductive capacity; but formulæ of extrapolation are always dangerous when used far from the actual observations. If Maxwell's theory is true, light flint-glass should be perfectly transparent to radiations having a wave-period of, let us say, $\frac{1}{17000}$ second; because this glass is sensibly a perfect electrical insulator, its refractive index for such waves should be about 2·6. Are there any facts to induce us to think such a thing possible? It is well known that in some cases strong selective absorption of light in the visible spectrum causes what is known as anomalous dispersion; that is to say, the body which presents such selective absorption of certain rays has a refractive index abnormally low for waves a little shorter than those absorbed, and an index abnormally high for waves a little longer than those absorbed*.

Light flint-glass is very transparent through the whole visible spectrum, but it is by no means transparent in the infra-red. If the absorption in the infra-red causes in light flint-glass anomalous dispersion, we should find a diminished refractive index in the red. We may say that we have a hint of this; for if we represent the refractive indices by the ordinates of a curve in which the squares of the reciprocals of the wave-lengths are abscissæ, this curve presents a point of inflection†. In the part corresponding to short waves it is concave upwards; in the part corresponding to long waves it is concave downwards: the curvature, however, is very slight. Does it not seem possible, looking at the matter from the purely optical point of view, that if we could examine the spectrum below the absorption in the infra-red, we should find the effect of anomalous dispersion, and that the

---

* *Theory of Sound*, by Lord Rayleigh, vol. I. p. 125.

† *Proceedings of the Royal Society*, 1877.

refractive index of such long waves might even be so high as 2·6 ? To test this experimentally in a conclusive manner would probably not be easy.   Perhaps the best chance of finding how these long waves are refracted would be to experiment on the rays from a thermopile to a freezing-mixture.   Without an actual measurement of a refractive index below all strong absorption, it cannot be said that experiment is in contradiction to the Electromagnetic Theory of Light ; for a strong absorption introduces a discontinuity into the spectrum which forbids us from using results on one side of that discontinuity to infer what they would be on the other side.

24.

ON THE QUADRANT-ELECTROMETER*.

[From the *Philosophical Magazine*, April, 1885, pp. 291—303.]

In Professor Clerk Maxwell's *Electricity* (vol. I. p. 273) it is proved that the deflection of the needle of a quadrant-electrometer varies as $(A - B)\left(C - \dfrac{A + B}{2}\right)$, where C is the potential of the needle, and A and B of the two pairs of quadrants. Desiring to ascertain the value of the standard charge of my instrument, I endeavoured to do so by the aid of this formula, and also by a more direct method. The results were quite discordant. Setting aside the special reasoning by which the formula is obtained, we should confidently expect that the sensibility of a quadrant-electrometer would increase continuously as the charge of the jar is increased, until at last a disruptive discharge occurs. In my instrument this is not the fact. As the charge was steadily increased by means of the replenisher, the deflection of the needle due to three Daniell's elements at first increased, then attained a maximum, and with further increase of charge actually diminished. On turning the replenisher in the inverse direction the sensibility at first increased, attained the maximum previously observed, and only on further reduction of charge diminished.

Before giving the experimental results, it may be worth while to briefly examine the theory of the quadrant-electrometer. Let A, B, C, D be the potentials of the quadrants, the needle, and the

* Read before the Physical Society on March 14th, 1885.

inductor which is used for measuring high potentials (see Reprint of Sir W. Thomson's papers, p. 278). Let Q_1, Q_2, Q_3, Q_4 be the quantities of electricity on these bodies respectively, and θ the angle of deflection of the needle, measured in terms of divisions of the scale, on which the image of the lamp-flame is projected. We have the equations

$$\left.\begin{aligned}
Q_1 &= q_{11}A - q_{12}B - q_{13}C - q_{14}D \\
Q_2 &= -q_{12}A + q_{22}B - q_{23}C - q_{24}D \\
Q_3 &= -q_{13}A - q_{23}B + q_{33}C - q_{34}D \\
Q_4 &= -q_{14}A - q_{24}B - q_{34}C + q_{44}D
\end{aligned}\right\} \quad \dots\dots\dots\dots(1).$$

q_{11} &c. are the coefficients of capacity and induction. They are independent of A, B, C, D, and are functions of θ only. As above written, they are all positive. Let the energy of electrification be W:—

$$\left.\begin{aligned}
2W = &q_{11}A^2 + q_{22}B^2 + q_{33}C^2 + q_{44}D^2 \\
&- 2q_{12}AB - 2q_{13}AC - 2q_{14}AD \\
&- 2q_{23}BC - 2q_{24}BD \\
&- 2q_{34}CD
\end{aligned}\right\} \quad \dots\dots\dots\dots(2).$$

Equations (1) and (2) are perfectly general, true whatever be the form of the four bodies.

If the four quadrants completely surround the needle,

$$\left.\begin{aligned}
&q_{34} = 0 \\
&q_{14}, \; q_{24}, \text{ and } q_{44} \text{ are independent of } \theta \\
&q_{33} = q_{13} + q_{23}
\end{aligned}\right\} \quad \dots\dots\dots(3).$$

Now when the electrometer is properly adjusted, the needle will not be deflected when $A = B$, whatever C and A may be. Hence $A - B$ is a factor of $\dfrac{dW}{d\theta}$, and we have

$$\left.\begin{aligned}
&\frac{dq_{11}}{d\theta} + \frac{dq_{22}}{d\theta} = 2\,\frac{dq_{12}}{d\theta} \\
&\frac{dq_{33}}{d\theta} = 0
\end{aligned}\right\} ;$$

whence

$$2\frac{dW}{d\theta} = (A - B)\left(A\,\frac{dq_{11}}{d\theta} - B\,\frac{dq_{22}}{d\theta} - 2C\,\frac{dq_{13}}{d\theta} \right).$$

This should be true of any electrometer having the above adjustment correctly made.

But by suitably forming the three bodies A, B, C, further relations between the coefficients may be obtained. The condition of symmetry would give us $\dfrac{dq_{11}}{d\theta} = -\dfrac{dq_{22}}{d\theta}$; but it is not necessary to assume symmetry. If the circumferential termination of the needle be a circle centre in the axis of suspension (at least near the division of the quadrants), if the needle turn in its own plane, if the quadrants are each approximately a surface of revolution about the axis, and if the radial terminations of the needle be not within the electrical influence of the quadrants within which they are not, conditions closely satisfied in Sir W. Thomson's electrometer,

$$\frac{dq_{11}}{d\theta} = \frac{dq_{13}}{d\theta},$$

$$\frac{dq_{22}}{d\theta} = -\frac{dq_{13}}{d\theta}.$$

If θ be small, we obtain

$$\frac{dW}{d\theta} = \alpha\,(A - B)\left(C - \frac{A + B}{2}\right),$$

the formula in Maxwell.

Returning now to our original equation, we have

$$
\begin{aligned}
2W = \;& q_{11}A^2 + q_{22}B^2 + q_{33}C^2 + q_{44}D^2 \\
& - 2q_{12}AB - q_{33}AC - 2q_{14}AD \\
& - q_{33}BC - 2q_{24}BD \\
& + 2\alpha\theta\,(A - B)\left(C - \frac{A + B}{2}\right);
\end{aligned}
$$

involving in all eight constants, q_{11} &c. being now regarded as representing the values of the coefficients in the zero position.

$$
\left.
\begin{aligned}
Q_1 &= q_{11}A - q_{12}B - \tfrac{1}{2}q_{33}C - q_{14}D + \alpha\theta\,(C - A) \\
Q_2 &= - q_{12}A + q_{22}B - \tfrac{1}{2}q_{33}C - q_{24}D - \alpha\theta\,(C - B) \\
Q_3 &= -\tfrac{1}{2}q_{33}A - \tfrac{1}{2}q_{33}B + q_{33}C \phantom{- q_{24}D} + \alpha\theta\,(A - B) \\
Q_4 &= - q_{14}A - q_{24}B \phantom{- \tfrac{1}{2}q_{33}C} + q_{44}D
\end{aligned}
\right\}.
$$

We may now discuss a variety of important particular cases.

(*a*)   *B* is put to earth ; *A* then is connected to a condenser, capacity *a*, charged to potential *V* : we want to know *V* from the reading of the electrometer.   Here

$$aV - \tfrac{1}{2}q_{33}C = (a + q_{11})A - \tfrac{1}{2}q_{33}C + \alpha\theta\,(C - A),$$

$$V = A + \frac{q_{11}}{a}A + \frac{\alpha\theta\,(C - A)}{a}.$$

Neglecting *A* compared with *C*, and assuming

$$\theta = \lambda\,(A - B)\left(C - \frac{A + B}{2}\right),$$

we have

$$V = A\left\{1 + \frac{q_{11}}{a} + \frac{\lambda\alpha C^{2}}{a}\right\}.$$

The apparent capacity of *A* increases with *C*.

(*b*)   *B* is again zero.   *A* is connected to a source, but is disconnected and insulated when the deflection of the needle is θ' ; the final deflection is θ : required the potential *V* of the source.

$$q_{11}V - \tfrac{1}{2}q_{33}C + \alpha\theta'\,(C - V)$$

$$= q_{11}A - \tfrac{1}{2}q_{33}C + \alpha\theta\,(C - A),$$

$$V = A + \frac{\alpha\,(\theta - \theta')\,C}{q_{11}}$$

$$= A\left\{1 + \frac{\alpha\lambda\left(1 - \dfrac{\theta'}{\theta}\right)C^{2}}{q_{11}}\right\}.$$

We may now consider the methods of varying the sensibility of the instrument (see Reprint of Sir W. Thomson's papers, p. 280).   The methods dealt with are those of Sir W. Thomson, somewhat generalized.

(*c*)   The quadrant *B* is connected with an insulated condenser, capacity *b*, whilst *A* is connected to a source of electricity :—

$$0 = -q_{12}A + (b + q_{22})B - \alpha\theta C,$$

$$\theta = \lambda\,(A - B)\,C ;$$

therefore

$$0 = -q_{12}A + (b + q_{22})\left(A - \frac{\theta}{\lambda C}\right) - \alpha\theta C ;$$

so

$$A = \theta\,\frac{\dfrac{b + q_{22}}{\lambda C} + \alpha C}{(b + q_{22}) - q_{12}}.$$

If $b = 0$, we have the first reduced sensibility given by Sir W. Thomson.

(d) All methods of using the inductor may be treated under one general form. Let the quadrants A and B be connected with insulated condensers, capacities a and b; then connect the inductor to a source, potential V;

$$\left. \begin{aligned} 0 &= (q_{11} + a)\, A - q_{12} B - q_{14} V + \alpha \theta C \\ 0 &= -\, q_{12} A + (q_{22} + b)\, B - q_{24} V - \alpha \theta C \\ \theta &= \lambda\, (A - B)\, C \end{aligned} \right\};$$

$$(+\, q_{11} q_{22} - q_{12}{}^2 + a q_{22} + b q_{11} + ab)\, (A - B)$$
$$+ \{q_{14}\, (q_{12} - q_{22} - b) - q_{24}\, (q_{12} - q_{11} - a)\}\; V$$
$$+ \{-\, 2 q_{12} + q_{22} + q_{11} + b + a\}\, \alpha \theta C = 0\,;$$

whence we have an expression for V proportional to θ. By a proper choice of a and b, we can make the sensibility as low as we please.

Now the whole of these formulæ rest on the same reasoning as the equation

$$\theta = \lambda\, (A - B) \left(C - \frac{A + B}{2} \right).$$

I have mentioned that, in my instrument at least, this equation quite fails to represent the facts when C is considerable. It becomes a matter of interest to ascertain when the formula begins to err to a sensible extent. If a constant battery of a large number of elements were available, this would be soon accomplished. I have at present set up only 18 Daniells. I have therefore been content to use the electrometer to ascertain its own charge by the aid of the inductor, using the 18 Daniells as a standard potential. As the charges range as high as 2600 Daniell's elements, the higher numbers can only be regarded as very rough approximations; sufficiently near, however, to indicate the sort of result which would be obtained if more precise methods were used. The first column in the following Table gives the ascertained or estimated charge of the jar of my electrometer in Daniell's elements; the second the deflection in scale-divisions caused by three elements; the third, the coefficient λ, deduced by the formula $\theta = \lambda A C$: this coefficient ought theoretically to be constant.

I.	II.	III.
72	75	0·35
112	118	0·35
136	140	0·35
178	190	0·35
238	239	0·34
303	288	0·32
383	336	0·30
512	391	0·26
616	409	0·22
813	432	0·18
1080	424	0·13
1312	402	0·10
1728	360	0·07
2124	320	0·05
2634	296	0·037
1704	353	0·07
1436	394	0·09
1284	412	0·11
876	436	0·17
684	427	0·21

By connecting the jar and one quadrant to 18 elements and the other quadrant to earth, I obtained 0·356 as the value of λ, making use of the complete equation

$$\theta = \lambda (A - B)\left(C - \frac{A + B}{2}\right).$$

It will be seen that this equation may be trusted until C is over 200 Daniell's elements potential, but that when C exceeds 250 a quite different law rules.

The foregoing was read before the Physical Society a few years ago, but I stopped its publication after the type was set up, because I was not satisfied that my appliances for experiment were satisfactory, or that I could give any satisfactory explanation of the anomaly.

The electrometer had been many times adjusted for various purposes before further experiments were made, so that those which I shall now describe cannot be directly compared with what goes before.   The old experiment was first repeated, and the existence of a maximum sensibility again found. On examination, it was found that the needle hung a little low so that it was nearer to the part of the quadrant below it than to that above. It is easy to see that this would produce the anomalous result observed, though there is reason for thinking it is not the sole cause. The effect of the needle being low is that it will be on the whole attracted downwards; and so the apparent weight hanging on the fibre-suspension and the consequent tension of the fibres will be increased. The increase of the tension will be as the square of the potential C; and hence the formula for the deflection will be modified to

$$\theta = \frac{\lambda}{1 + kC^2}(A - B)\left(C - \frac{A + B}{2}\right),$$

where k is a constant depending upon the extent to which the position of the needle deviates from its true position of midway between the upper and lower parts of the quadrants. By a proper choice of k, the results I previously obtained are found to agree well with this formula.

The electrometer was next adjusted in the following way:— The needle was raised by taking up the fibres of the suspension and adjusting them to equal tension in the usual way, and the proportionality of sensibility to charge was tested, the charge being now determined in arbitrary units by discharging the jar of the instrument through a ballistic galvanometer. The operation was repeated until the sensibility, so far as this method of testing goes, was proportional to the charge of the jar over a very long range. It was then found that the needle was slightly above the median position within the quadrants. Increased tension of the fibres from electrical attraction does not therefore account for the whole of the facts, although it does play the principal part. The sensibility of the instrument being now at least approximately proportional to the charge of the jar, I proceeded to determine accurately the potential of the jar when charged to the standard as indicated by the idiostatic gauge.

In what follows the quadrants, one of which is under the induction-plate, are denoted by B, the others by A. The quadrants B are connected to the case, A are insulated. The jar is connected to the induction-plate, and the reading on the scale noted; the connexion is broken, and the induction-plate is connected to the case, and the reading on the scale again noted; the difference is the deflection due to the charge in the jar. It is necessary to read the scale for zero-charge on induction-plate last, because the charging of the induction-plate slightly diminishes the charge of the jar, and considerably displaces the zero-reading by giving an inductive charge to the quadrant A. It is also necessary to begin with the charge of the jar minutely too high, so that after separating the induction-plate from the interior of the jar, the latter shall have exactly the correct charge as indicated by the gauge. The deflection thus obtained was precisely $298\frac{1}{2}$, repeated in many experiments. The *double* deflection given by seventy Daniell cells was 43·6 scale-divisions. By comparison with two Clark's cells, the value of which I know, the potential of the seventy Daniells was found to be 74·2 volts; hence the potential of the jar is 1016 volts, when charged to the potential indicated by the gauge.

The constant λ of the instrument was next determined by the formula

$$\theta = \lambda (A - B) \left(C - \frac{A + B}{2} \right).$$

Four modes of connecting are available for this :—

$$A = C = 74·2 \text{ volts, } B = 0;$$
$$B = C = 74·2 \text{ volts, } A = 0;$$
$$A = C = 0, B = 74·2 \text{ volts};$$
$$B = C = 0, A = 74·2 \text{ volts}.$$

In each case the deflection was 253·5 if the charge on the needle was positive in relation to the quadrant with which it was not connected; and was 247 when the needle was negative. This at first appeared anomalous; but the explanation is very simple. The needle is aluminium, the quadrants are either brass or brass-gilded, I am not sure which. There is therefore a contact-difference of potential between the needle and the quadrants; call it x. Thus, instead of $\theta = \frac{A^2}{2}$, we have

$$\theta = + \lambda A \left(\frac{A}{2} + x \right)$$

and

$$\theta = \lambda (- A) \left(- \frac{A}{2} + x \right);$$

this gives $\qquad x = \dfrac{6\cdot5}{2\lambda A} = 0\cdot482$ volt.

The result was verified by using fourteen cells instead of seventy: the deflections were $10\cdot0$ and $8\cdot8$, which gives the same value to x. It is worth noting that the same cause affects the idiostatic gauge in the same way. Let the jar be charged till the gauge comes to the mark. Call P the difference of potential between the aluminium lever of the idiostatic gauge and the brass disk below which attracts it. The difference of potential between the brass of the case and the brass work of the interior is $P + x$, and between the case and the aluminium needle within the quadrants it is $P + 2x$. If, however, the charge is negative, the difference is $-P + 2x$. Hence the sensibility will be different from two causes, according as the jar is charged positively or negatively, till the idiostatic gauge is at its standard. For determining the constant λ we must take the mean of the two results $253\cdot5$ and 247, that is $250\cdot25$. Comparing this with the actual standard charge of the jar, and the double deflection given by one volt $172\cdot4$, when charged to the standard, we see that the irregularity has not been wholly eliminated. It appeared desirable to determine the sensibility of the instrument for a lower known charge. The charge was determined exactly as described above and was found to be 609 volts; whilst 1 volt gave $107\cdot1$ scale-divisions double deflection; whence in the equation

$$\theta = \frac{\lambda(A - B)\,C}{1 + kC^2},$$

we have, if
$$\lambda = 0\cdot1816, \quad k = 7 \times 10^{-8},$$

the following as the calculated and observed deflections:—

Calculated . . . . $250\cdot0$, $107\cdot7$, $172\cdot4$,

Observed . . . . $250\cdot2$, $107\cdot1$, $172\cdot4$,

which is well within errors of observation.

This deviation from proportionality of sensibility did not appear to be worth correcting, as I was not sure that other small irregularities might not be introduced by raising the needle above the middle position within the quadrants. It appears probable that the small deviation still remaining does not arise from the attraction of the quadrants on the needle increasing the

tension of the suspension, but from some cause of a quite different
nature, for if it were so caused the capacity-equations would be

$$Q_1 = q_{11}A - q_{12}B - q_{33}C - q_{14}D + \alpha\theta C, \text{ &c.,}$$

where $$\theta = \frac{\lambda}{1 + kC^2}(A - B)\left(C - \frac{A + B}{2}\right)^*.$$

Now the experiments I have tried for determining q_{11}, q_{12}, &c., are
not in accord within the limits of errors of observation, using these
equations of capacity; but they are in better accord if, in lieu of
the term $\alpha\theta C$, we write $\dfrac{\alpha\theta C}{1 + kC^2}$. I have no explanation of this
to offer; but in what follows it is assumed that the equations
expressing the facts are

$$\left.\begin{aligned}
\theta &= \mu(A - B), \text{ where } \mu = \frac{\lambda C}{1 + kC^2} \\
Q_1 &= q_{11}A - q_{12}B - q_{14}D + \beta\mu\theta \\
Q_2 &= -q_{12}A + q_{22}B - q_{24}D - \beta\mu\theta \\
Q_4 &= -q_{14}A - q_{24}B + q_{44}D
\end{aligned}\right\}.$$

We are now in a position to determine the various coefficients
of capacity: in doing so it is necessary to distinguish the values
of q_{11} and q_{22} when the posts by which contact with the quadrants
is made are down and in contact with the quadrants, and when
they are raised up out of contact; the former are denoted by
$q_{11} + a$ and $q_{22} + a$, the latter by q_{11} and q_{22}, the capacity of the
binding-posts being a. As a convenient temporary unit of capacity
the value of $\beta\mu^2$, when the jar has the standard charge, is taken.
The first set of experiments was to determine the deflections
caused by known potentials with varied charge of jar, one or
other of the quadrants being insulated. Three potentials of the
jar were used—that of the standard indicated by the idiostatic
gauge and two lower. The values of μ are denoted by μ_3, μ_2, μ_1.
It was found by connecting the two quadrants to standard cells
that

$$\mu_3 : \mu_2 : \mu_1 = 1 : 0.805 : 0.585;$$

and hence

$$\beta\mu_3{}^2 = 1, \quad \beta\mu_2{}^2 = 0.648, \quad \beta\mu_1{}^2 = 0.342.$$

Suppose quadrant A be insulated, and potential B be applied to
quadrant B; then we have, if θ be the deflection which potential B

---

* The cause of this was determined by Messrs Ayrton, Perry, and Sumpner to
lie in the shape of the guard-tube. *Phil. Trans. R. S.* Vol. 182, p. 539.

would cause with standard charge, if quadrant A were connected to the case, and ϕ the observed deflection,

$$0 = q_{11}A - q_{12}B + \beta\mu\phi \,;$$
$$\phi = \mu\,(A - B)\,;$$
$$\theta = - \mu_3 B \,;$$

whence

$$\phi = \theta \cdot \frac{\mu}{\mu_3} \cdot \frac{q_{11} - q_{12}}{q_{11} + \beta\mu^2}.$$

In the calculated values of ϕ given below,

$$q_{11} = 0\text{·}502, \qquad q_{22} = 0\text{·}543,$$
$$q_{12} = 0\text{·}293, \qquad a = 0\text{·}200 \text{ for } B,$$
$$= 0\text{·}193 \text{ for } A.$$

A closer approximation to observation is obtained by assuming the two contact-posts to be of slightly different capacities; the difference given above is no more than might be expected to exist.

The jar being charged to standard potential, B was insulated and its post raised, and A was connected to 10 Daniells, for which $\theta = 1808$:—

> Deflection observed = 293·2,
> „ calculated = 293·0.

The post of B was lowered to contact :—

> Deflection observed = 467·0,
> „ calculated = 466·8.

A was now insulated and post raised, B was connected to the same battery :—

> Deflection observed = 251·0,
> „ calculated = 251·6.

The post of A was lowered to contact :—

> Deflection observed = 429·0,
> „ calculated = 428·8.

The jar was now charged to a lower potential, for which $\mu = \mu_2$, with B insulated and post raised, and A connected to 30 Daniells, for which $\theta = 5468$:—

> Deflection observed = 925·0,
> „ calculated = 924·0.

The post of B was lowered to contact, and A connected to 10 Daniells, for which $\theta = 1808$:—

$$\text{Deflection observed} = 470\cdot5,$$
$$\text{,,} \quad \text{calculated} = 470\cdot85.$$

A was now insulated and post raised, B was connected to a battery of 30 Daniells, for which $\theta = 5468$:—

$$\text{Deflection observed} = 798\cdot0,$$
$$\text{,,} \quad \text{calculated} = 800\cdot0.$$

The post of A was lowered to contact, and B was connected to 10 Daniells; $\theta = 1808$:—

$$\text{Deflection observed} = 437\cdot0,$$
$$\text{,,} \quad \text{calculated} = 435\cdot7.$$

The jar was then charged to a still lower potential, for which $\mu = \mu_1$, with B insulated and post raised, and A connected to 30 Daniells, for which $\theta = 5468$:—

$$\text{Deflection observed} = 901\cdot0,$$
$$\text{,,} \quad \text{calculated} = 903\cdot6.$$

The post of B was lowered to contact and A connected to 10 Daniells; $\theta = 1808$:—

$$\text{Deflection observed} = 437\cdot0,$$
$$\text{,,} \quad \text{calculated} = 438\cdot7.$$

A was now insulated and post raised, and B was connected to 30 Daniells; $\theta = 5468$:—

$$\text{Deflection observed} = 785,$$
$$\text{,,} \quad \text{calculated} = 792.$$

The post of A was lowered to contact and B connected to 10 Daniells; $\theta = 1808$:—

$$\text{Deflection observed} = 408,$$
$$\text{,,} \quad \text{calculated} = 410.$$

The next experiment was similar, excepting only that the insulated quadrant B was connected to a condenser; this condenser consisted merely of a brass tube insulated within a larger tube—its capacity is about $0\cdot00009$ microfarad. The jar was at its standard charge. Calling the capacity of the condenser b, in terms of our temporary unit, we have, as before,

$$\phi = \theta \frac{b + a + q_{22} - q_{12}}{b + a + q_{22} + 1} \ [\beta \mu_3{}^2 = 1].$$

When $\theta = 1259$, ϕ was observed to be 927, whence $b = 3\cdot159$.

We are now in a position to obtain independent verification of the values already obtained for the constants. Suppose A be connected to the case, that condenser b is charged from a battery of known potential, such that it would give deflection θ if connected to B, and the charged condenser is then connected to B. Suppose ψ be the deflection before connexion is made, ϕ after. Then

$$b\,(\theta - \phi) = \{q_{22} + a + 1\}\,(\phi - \psi).$$

When $\theta = 1439$ and $\psi = 0$, it was found that $\phi = 915$. The value of ϕ, calculated from the values of the constants already obtained, is 928.

When $\theta = 1439$ and $\psi = -676$, it was found that $\phi = +676$; the calculated value is 688.

A further experiment of verification, involving only the capacity of the quadrant, is the following. The quadrant A being connected to the case, B was charged by contact instantaneously made and broken with a battery of known potential, and the resulting deflection was noted. The instantaneous contact being made by hand, no very great accuracy could be expected. Let ψ and ϕ be the readings on the scale before and after the instantaneous contact; then

$$\frac{\theta - \phi}{\phi - \psi} = \frac{1}{q_{22} + a} = 1\cdot345.$$

The following results were obtained :—

θ.	ψ.	ϕ observed.	ϕ calculated.
1796	0	763	765
1796	-493	493	482

We next determine the coefficients q_{14} and q_{24} of induction of the induction-plate on the quadrants. This is easily done from the deflections obtained with the induction-plate, one or both pairs of quadrants being insulated. First, suppose one pair, say B, are insulated whilst A is connected to the case :—

$$0 = + q_{22}\, B - q_{24}\, D - \beta\mu\phi,$$
$$\phi = - \mu B,$$
$$\theta = \mu_3 D\,;$$

whence

$$\phi = \theta\, \frac{\mu}{\mu_3}\, \frac{q_{24}}{q_{22} + \beta\mu^2}\,,$$

ϕ being the deflection actually observed, and θ that which the battery used would give if connected direct to the quadrants, the needle having the standard charge. When θ was 12,800 and $\mu = \mu_3$, ϕ was 418, whence $q_{24} = 0\cdot0504$.

In the same way, A being insulated but B connected to the case, ϕ was found to be 43·6, whence $q_{14} = 0\cdot00508$.

Again, when both quadrants are insulated we have

$$0 = \quad q_{11}A - q_{12}B - q_{14}D + \beta\mu\phi,$$
$$0 = - q_{12}A + q_{22}B - q_{24}D - \beta\mu\phi,$$
$$\phi = \mu\,(A - B),$$
$$\theta = \mu_3 D.$$

From the first two equations,

$$(q_{11}q_{22} - q_{12}{}^2)\,(A - B) - \{(q_{22} - q_{12})\,q_{14} - (q_{11} - q_{12})\,q_{24}\}\,D$$
$$+ (q_{22} + q_{11} - 2q_{12})\,\beta\mu\phi = 0\,;$$

whence

$$\phi = \theta\, \frac{\mu}{\mu_3}\, \frac{(q_{22} - q_{12})\,q_{14} - (q_{11} - q_{12})\,q_{24}}{(q_{11}q_{22} - q_{12}{}^2) + (q_{22} + q_{11} - 2q_{12})\,\beta\mu^2}.$$

In the case when $\mu = \mu_3$, substituting the values already determined, we have

$$\phi = \theta \times 0\cdot0142\,;$$

it was observed with $\theta = 12,800$ that $\phi = 183$; the calculated value would be 182.

With a lower charge on the jar, viz. when $\mu = \mu_3 \times 0\cdot805$, with B insulated, A connected to the case, and $\theta = 12,800$, it was found that $\phi = 437\cdot5$; the calculated value is 441.

The capacity q_{44} of the induction-plate is of no use; its value, however, is about $0\cdot004$, in the same unit as has been so far used.

The capacity q_{33} of the needle and the coefficient of induction of the needle on either quadrant $\tfrac{1}{2}q_{33}$ are also of no use, but the method by which they may be obtained is worth noting. Let quadrants A be connected to the case, and let B be insulated, diminish the charge of the jar slightly by the replenisher, and suppose the consequent deflection be ϕ. Let μ and μ' be the values of μ before and after the diminution of charge, as ascertained by applying a known potential-difference between the two pairs of quadrants; we have

$$- q_{33} C = + q_{22} B - q_{33} C' - \beta \mu' \phi,$$

where

$$\phi = - \mu' B,$$
$$q_{33} (C - C') = (q_{22} + \beta \mu'^2) B,$$

which determine q_{33}, since C and C' are known from μ and μ'.

Of course the values of the constants of an electrometer are of no value for any instrument except that for which they are determined in the state of adjustment at the time. For any particular use of the instrument it is best to determine exactly that combination of constants which will be needed. Nor is there anything new in principle in the discussion or experiments here given; they are merely for the most part the application of well-known principles to methods of using the electrometer given by Sir William Thomson himself. The method of determining the capacity of a condenser by charging it and connecting it to an insulated quadrant has been used by Boltzmann. But the invention of the quadrant-electrometer by Sir William Thomson may be said to have marked an epoch in Electrostatics, and the instrument from time to time finds new uses. It therefore seems well worth while to make known observations made upon it in which the instrument itself has been the only object studied. Some practical conclusions may, however, be drawn from the preceding experiments. Before using the formula

$$\theta = \lambda (A - B) \left(C - \frac{A + B}{2} \right)$$

it is necessary to verify that it is sufficiently nearly true, or to determine its variation from accuracy. Unless it be sufficiently accurate through the range experimented upon, the electrometer cannot be applied by the methods well known for determining alternating potentials and the work done by alternating currents.

My pupil, Mr Paul Dimier, has very efficiently helped me in the execution of the experiments of verification.

25.

NOTE ON SPECIFIC INDUCTIVE CAPACITY.

[From the *Proceedings of the Royal Society*, Vol. XLI., pp. 453—459.]

Received November 9, 1886.

CONSIDER a condenser formed of two parallel plates at distance x from each other, their area A being so great, or the distance x so small, that the whole of the lines of force may be considered to be uniformly distributed perpendicular to the plates. The space between the plates is occupied by air, or by any insulating fluid. Let e be the charge of the condenser and V the difference of potential between the plates. If the dielectric be air, there is every reason to believe that $V \propto e$, that is, there is for air a constant of specific inductive capacity. My own experiments ([1881] *Phil. Trans.*, vol. CLXXII. p. 355) show that in the case of flint-glass the ratio of V to e is sensibly constant over a range of values of V from 200 volts per cm. to 50,000 volts per cm. From experiments in which the dielectric is one or other of a number of fluids and values of V upwards of 30,000 volts per cm. are used, Professor Quincke concludes (*Wiedemann's Annalen*, vol. XXVIII., 1886, p. 549) that the value of e/V is somewhat less for great electric forces than for small. From the experiments described in that paper, and from his previous experiments (*Wiedemann's Annalen*, vol. XIX., 1883, p. 705, *et seq.*) he also concludes that the specific inductive capacity determined from the mechanical force resisting separation of the plates is 10 per cent. to 50 per cent. greater than that determined by the actual charge of the condenser. The purpose of the present note is to examine the relations of these important conclusions, making as few assumptions as possible.

The potential difference V is a function of the charge e and distance x, and if the dielectric be given of nothing else. The work done in charging the condenser with charge e is $\int_0^e V\,de$. If the distance of the plates be changed to $x + dx$, the work done in giving the same charge is $\int_0^e \left(V + \dfrac{dV}{dx}\,dx\right)de$, hence the mechanical force resisting separation of the plates is $\int_0^e \dfrac{dV}{dx}\,de$. If the dielectric be air, $A\,\dfrac{V}{x} = 4\pi e$, and the attractive force between the plates is $\dfrac{2\pi e^2}{A}$ or $\dfrac{A V^2}{8\pi x^2}$. If K_p be the dielectric constant as determined by an experiment on the force between the plates when the potential difference is V and distance is x,

$$K_p = \int_0^e \frac{dV}{dx}\,de \left/ \frac{A V^2}{8\pi x^2} \right. \quad\ldots\ldots\ldots\ldots\ldots(1).$$

If K be the dielectric constant obtained by direct comparisons of charge and potential,

$$K = \frac{4\pi x e}{A V} \quad\ldots\ldots\ldots\ldots\ldots\ldots(2),$$

whence

$$K_p/K = \int_0^e \frac{dV}{dx}\,de \left/ \frac{V e}{2x} \right. \quad\ldots\ldots\ldots\ldots\ldots(3).$$

We ordinarily assume that $V \propto x e$; if so, $K_p/K = 1$. These results follow quite independently of any suppositions about the nature of electricity, about action at a distance, or tensions and pressures in the dielectric.

Yet another method of determining the dielectric capacity of fluids has been used by Professor Quincke. Let a bubble of air be introduced between the two plates, let the area of the bubble be A_1, and let P be the excess of pressure in the bubble above that in the external air when the potential is V, allowance being first made for capillary action.

The condenser now consists of two parts, one a fluid condenser area $A - A_1$, the other an air condenser area A_1; we have mechanical work done in increasing the area of the bubble from A_1 to $A_1 + dA_1$, with constant charge—

$$e = \int_0^e \frac{dV}{dA_1} dA_1 de;$$

but this work is

$$xPdA_1,$$

whence

$$xP = \int_0^e \frac{dV}{dA_1} de.$$

Now

$$4\pi e = A_1 \frac{V}{x} + (A - A_1) f(V),$$

where $4\pi e = Af(V)$, when the whole space is occupied by fluid, and the distance is x.

The charge being constant we have—

$$0 = \left\{ \frac{V}{x} - f(V) \right\} dA_1 + \left\{ \frac{A_1}{x} + (A - A_1) f'(V) \right\} dV,$$

and for the purpose of transforming the integral

$$4\pi de = \left\{ \frac{A_1}{x} + (A - A_1) f'(V) \right\} dV,$$

$$4\pi \frac{dV}{dA_1} de = \left\{ f(V) - \frac{V}{x} \right\} dV,$$

whence

$$xP = \frac{1}{4\pi} \int_0^V \left\{ f(V) - \frac{V}{x} \right\} dV \dots\dots\dots\dots(4),$$

$$x \frac{dP}{dV} = \frac{1}{4\pi} \left\{ f(V) - \frac{V}{x} \right\} \quad \dots\dots\dots\dots(5).$$

Writing with Quincke K_s for the dielectric constant determined by a measurement of P, we have by substituting in (4)

$$f(V) = \frac{K_s V}{x},$$

and integrating as though K_s were constant,

$$xP = \frac{1}{4\pi} \cdot \frac{K_s - 1}{x} \cdot \frac{V^2}{2},$$

$$K_s = 1 + \frac{8\pi x^2 P}{V^2} \dots\dots\dots\dots\dots\dots(6),$$

which may be taken as the definition of K_s, whence

$$K_s = \frac{2x^2}{V^2} \int_0^V \frac{f(V)}{x} dV \dots\dots\dots\dots\dots\dots\dots\dots\dots (7);$$

but from (5) we have, since in fact $K = \frac{xf(V)}{V}$,

$$x \frac{dP}{dV} = \frac{1}{4\pi} \{K - 1\} \frac{V^2}{x} \dots\dots\dots\dots\dots\dots\dots (8).$$

But

$$K_p = \int_0^e \frac{dV}{dx} de \left| \frac{AV^2}{8\pi x^2} \right.$$

$$= \int_0^V \frac{dV}{dx} f'(V) dV \left| \frac{V^2}{2x^2} \right.$$

$$= \frac{2x^2}{V^2} \left\{ f(V) \frac{dV}{dx} - \int_0^V f(V) \frac{d}{dV} \frac{dV}{dx} dV \right\}.$$

Hitherto we have made no assumption excepting that energy is not dissipated in a condenser by charge and discharge. We now make an assumption concerning $f(V)$, namely, that it is of the form $\phi(V/x)$, i.e., that $\frac{dV}{dx} = \frac{V}{x}$, or in words, that the capacity of a condenser varies inversely as the distance between the plates.

Then we have—

$$K_p = \frac{2x^2}{V^2} \left\{ f(V) \frac{V}{x} - \int_0^V \frac{f(V)}{x} dV \right\}$$

$$= 2K - K_s \dots\dots\dots\dots\dots\dots\dots\dots\dots\dots (9).$$

In words, the specific inductive capacity as determined by charge or discharge of a condenser at any given potential and distance between the plates is the arithmetic mean of the inductive capacity determined by the force resisting separation of the plates and of that determined by lateral pressure, the potential and distance being the same. This is true whatever be the relation between charge and potential difference, but it is at variance with the experimental result that K_p and K_s are both greater than K.

Further

$$\frac{K_p}{K} = \int_0^V Vf'(V) dV / \tfrac{1}{2} Vf(V).$$

In the accompanying curve, let abscissa of any point P of the curve OQP represent V, ordinate $f(V)$. If $K_p > K$ area $ONPQO$ > area of triangle ONP, i.e., unless the curve $y = f(x)$ has a point of inflection between O and P, the fact that $K_p > K$ implies that K increases with V,—a conclusion again at variance with experimental results.

We are thus unable to account for the observation on the hypothesis that the capacity varies inversely as x. Let us now suppose that $f(V) = V\phi(x)$, that is to say, that however the capacity may depend on the distance, it is independent of the

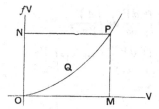

charge, or is constant for any given condenser. It at once follows that $K_s = K$, which is discordant with observation. Consider, however, the ratio—

$$K_p/K = \int_0^V \frac{dV}{dx} f'(V)\, dV \left/ \tfrac{1}{2} \frac{V}{x} f(V) \right. = -\frac{\dfrac{\phi'(x)}{\phi(x)}}{x},$$

since

$$0 = \frac{dV}{dx}\phi(x) + V\phi'(x),$$

when e is constant. Suppose $\dfrac{K_p}{K} = m$, a positive constant quantity greater than unity—

$$m\phi(x) + x\phi'(x) = 0,$$

$$x^m\phi(x) = C, \text{ a constant as regards } x,$$

or

$$\phi(x) \propto x^{-m}.$$

We could, therefore, account for K_p being greater than K by supposing that the potential difference with given charge per unit of area does not vary as x but as x^m. Such a supposition would be subversive of all accepted ideas of electrostatics.

There remains one other consideration to be named. We have assumed throughout that the charge of the condenser depends only on the distance of the plates and their difference of potential, and is independent of previous charges or of the time the difference of potential has existed. We have ignored residual charge. It is easy to see what its effect will be on determinations of K made by measuring the potential and charge of the condenser. It is not so obvious what its effect will be in all cases on the force between the plates. Consider a complete cycle of operations : the condenser is charged with quantity e, the distance between the plates is increased from x to $x + dx$, the condenser is discharged and the plates return to their initial position. The work done respectively in charging the condenser, separating the plates, and recovered in discharging the condenser, will depend on the rate at which these operations are performed. There are ideally two ways of performing them, so that no energy is dissipated by residual charge; first, under certain reservations, so rapidly that no residual charge is developed; second, so slowly that at each potential the residual charge is fully developed; in either case the potential is a function of the then charge, and not of the antecedent charges. The attraction between the plates will differ according as the charge is an instantaneous one or has been long applied. If a liquid were found exhibiting a considerable slowly-developed residual charge, the capacity determined by attraction with continuous charge would be greater than the capacity determined by an instantaneous discharge of the condenser through a galvanometer or into another condenser. I am not aware that residual charge has been observed in any liquid dielectric.

The results obtained by Professor Quincke are not easy to reconcile. For that reason it is the more desirable that their full significance should be ascertained. Full information is given of all the details of his experiments except on one point. It is not stated whether, in the experiments for determining K by direct discharge of the condenser, the capacity of the connexion and key was ascertained. It would in most ordinary arrangements of key be very appreciable in comparison with the capacity of the condenser itself. If neglected the effect would be to a certain extent to give too low a value of K, the effect being most marked when K is large.

I have made a few preliminary experiments to determine K for colza oil with several different samples, and both with continuous charges and intermittent charges from an induction coil. The values of K range from 2·95 to 3·11. Professor Quincke's results in his first paper are $K = 2\cdot443$, $K_p = 2\cdot385$, $K_s = 3\cdot296$.

The property of double refraction in liquids caused by electrification is sometimes cited as showing that electrification is not proportional to electromotive force. The fact that the double refraction in a liquid under powerful electromotive forces is very small would further show that there is a close approximation to proportionality, and that the deviation from proportionality would be insensible to any electrostatic test. Such conclusions, however, cannot be safely drawn in the case of bodies such as castor-oil, in which $K \neq \mu^2$. In such bodies, assuming the electromagnetic theory of light, the yielding to electromotive force is much greater if the force be applied for such time as 10^{-4} second than when applied for 10^{-14} second, and it is quite possible that the law of proportionality might be untrue in the former case, but very nearly or quite true in the latter.

ADDENDUM TO Dr HOPKINSON'S NOTE ON SPECIFIC INDUCTIVE CAPACITY. By PROFESSOR QUINCKE, For. Mem. R.S.

Received December 5, 1886.

Notiz über die Dielectricitätsconstante von Flüssigkeiten, von G. Quincke.

Bei Gelegenheit einer Untersuchung der Eigenschaften dielectrischer Flüssigkeiten (*Wiedemann's Annalen*, vol. XIX., 1883, p. 707; vol. XXVIII., 1886, p. 529) hatte ich die Dielectricitätsconstante mit der electrischen Wage oder dem hydrostatisch gemessenen Druck einer Luftblase grösser gefunden, als mit der Capacität eines Condensators, der von Luft oder isolirender Flüssigkeit umgeben ist, und beim Umlegen eines Schlüssels durch einen Multiplicator entladen wird.

Die Capacität des Schlüssels und des kurzen dünnen Verbindungsdrahtes, welcher den Schlüssel mit dem Condensator verband, wurde aber dabei als verschwindend klein vernachlässigt.

In Folge einer brieflichen Mittheilung von Herrn Dr John Hopkinson habe ich in neuster Zeit die Capacität des Schlüssels und des Zuleitungsdrahtes mit der Capacität C des Condensators durch Multiplicator-Ausschläge bei derselben Potentialdifferenz der Belegungen verglichen und dabei das Verhältniss—

$$\frac{x}{C} = 0 \cdot 1762$$

gefunden, also viel grösser als ich vermuthet hatte.

Zieht man von den beobachteten Multiplicator-Ausschlägen s_1 und s_{11} für den Condensator in Luft und in der dielectrischen Flüssigkeit den Ausschlag ab, der von der Electricität auf dem Schlüssel und Verbindungsdraht herrührt, so erhält man in der That durch das Verhältniss der so corrigirten Ausschläge (s_1) und (s_{11}) Werthe der Dielectricitätsconstante (K) der Flüssigkeit, die fast genau mit den Messungen der electrischen Wage übereinstimmen. Die Uebereinstimmung ist so gross, wie bei der Verschiedenheit der benutzten Beobachtungsmethoden nur erwartet werden kann.

So ergab sich z. B.

	Dielectricitätsconstante mit	
	Multipl. (K)	Wägung K_p
Aether	4·211	4·394
Schwefelkohlenstoff ...	2·508	2·623
„	2·640	2·541
Benzol	2·359	2·360
Steinöl	2·025	2·073

HEIDELBERG, *December* 1, 1886.

[Note added Dec. 4th.—Professor Quincke's explanation sets the questions I have raised at rest. There can be little doubt that K, K_s and K_p are sensibly equal and sensibly constant. The question what will happen to K_p and K_s if K is not constant has for the present a purely hypothetical interest.—J. H.]

26.

SPECIFIC INDUCTIVE CAPACITY.

[From the *Proceedings of the Royal Society*, Vol. XLIII., pp. 156—161.]

Received October 14, 1887.

THE experiments which are the subject of the present communication were originally undertaken with a view to ascertain whether or not various methods of determination would give the same values to the specific inductive capacities of dielectrics. The programme was subsequently narrowed, as there appeared to be no evidence of serious discrepancy by existing methods.

In most cases the method of experiment has been a modification of the method proposed by Professor Maxwell, and employed by Mr Gordon. The only vice in Mr Gordon's employment of that method was that plates of dielectrics of dimensions comparable with their thickness were regarded as of infinite area, and thus an error of unexpectedly great magnitude was introduced.

For determining the capacity of liquids, the apparatus consisted of a combination of four air condensers, with a fifth for containing the liquid arranged as in a Wheatstone's bridge, Fig. 1. Two, E, F, were of determinate and approximately equal capacity; the other two, J, I, were adjustable slides, the capacity of either condenser being varied by the sliding part. The outer coatings of the condensers E, F, were connected to the case of the quadrant electrometer, and to one pole of the induction coil; the outer coatings of the other pair, J, I, were connected to the needle of the electrometer and to the other pole of the induction coil. The inner coatings of the condensers J, F, were connected to one quadrant, and I, E, to the other quadrant of the electrometer. The slide of one or both condensers J, I, was adjusted till upon

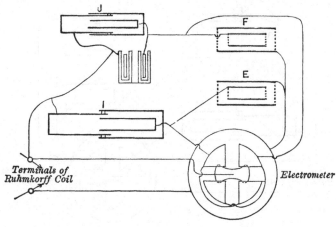

Fig. 1.

exciting the induction coil no deflection was observed on the electrometer. A dummy was provided with the fluid condenser, as in my former experiments, to represent the necessary supports and connexions outside of the liquid. Let now x be the reading of the sliding condenser when no condenser for fluid is introduced, and a balance is obtained. Let y be its reading when the condenser is introduced fitted with its dummy, z when the full condenser is charged with air. Let z_1 be the reading when the condenser charged with fluid is introduced, then will K, the specific inductive capacity of the liquid, be equal to $(y - z_1)/(y - z)$.

Three fluid condensers were employed, one was the same as in my former experiments[*]. Another was a smaller one of the same type arranged simply to contain a smaller quantity of fluid. The third was of a different type designed to prove that by no chance did anything depend on the type of condenser; this done it was laid aside as more complicated in use.

To determine the capacity of a solid, the guard-ring condenser of my previous experiments[†] was used. Advantage was taken of

* *Phil. Trans.* 1881, Part II.
† *Phil. Trans.* 1878, Part I.

the fact that at the time when there is a balance the potentials
of the interiors of all the condensers are the same.   Let the ring
O of the guard-ring condenser be in all cases connected to J, let
the inner plate of the guard-ring be connected to J as in Fig. 2,

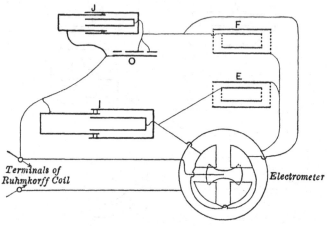

FIG. 2.

and let a balance be obtained.   Let the inner plate be now transferred to I as in Fig. 3, and again let a balance be obtained; the
difference of the two readings on the slide represents on a certain

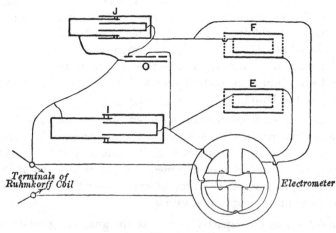

FIG. 3.

arbitrary scale the capacity of the guard-ring condenser at its then
distance.

In some cases it was necessary to adjust both condensers to obtain a balance, then the value of a movement of the scale of one condenser in terms of the other was known from previous experiment. In some cases it was found most convenient to introduce a condenser of capacity known in divisions of the scale of the sliding condenser coupled as forming part of the condenser *J*. The old method of adding the opposite charges of two condensers then connecting to the electrometer and adjusting until the electrometer remained undisturbed was occasionally used as a check; it was found to give substantially the same results as the method here described when the substance insulated sufficiently well to give any results at all.

Colza Oil. This oil had been found not to insulate sufficiently well for a test by the method of my former paper. Most samples, however, were sufficiently insulating for the present method. Seven samples were tested with the following mean results:—

No. 1. This oil was kindly procured direct from Italy for these experiments by Mr J. C. Field, and was tested as supplied to me—

$$K = 3\text{·}10.$$

No. 2 was purchased from Mr Sugg, and tested as supplied—

$$K = 3\text{·}14.$$

No. 3 was purchased from Messrs Griffin, and was dried over anhydrous copper sulphate—

$$K = 3\text{·}23.$$

No. 4 was refined rape oil purchased from Messrs Pinchin and Johnson, and tested as supplied—

$$K = 3\text{·}08.$$

No. 5 was the same oil as No. 4, but dried over anhydrous copper sulphate—

$$K = 3\text{·}07.$$

No. 6 was unrefined rape purchased from Messrs Pinchin and Johnson and tested as supplied, the insulation being bad, but still not so bad as to prevent testing—

$$K = 3\text{·}12.$$

No. 7. The same oil dried over sulphate of copper—

$$K = 3\cdot09.$$

Omitting No. 3, which I cannot indeed say of my own knowledge was pure colza oil at all, we may, I think, conclude that the
specific inductive capacity of colza oil lies between 3·07 and 3·14.

Professor Quincke gives 2·385 for the method of attraction
between the plates of a condenser, 3·296 for the method of lateral
compression of a bubble of gas.   Palaz* gives 3·027.

Olive Oil.   The sample was supplied me by Mr J. C. Field—

$$K = 3\cdot15.$$

The result I obtained by another method in 1880 was 3·16.

Two other oils were supplied to me by Mr J. C. Field.

Arachide. $K = 3\cdot17.$

Sesame. $K = 3\cdot17.$

A commercial sample of *raw linseed oil* gave $K = 3\cdot37.$

Two samples of *castor oil* were tried; one newly purchased
gave $K = 4\cdot82$; the other had been in the laboratory a long time,
and was dried over copper sulphate—

$$K = 4\cdot84.$$

The result of my earlier experiments for castor oil was 4·78;
the result obtained subsequently by Cohn and Arons† is 4·43.
Palaz gives 4·610.

Ether.   This substance as purchased, reputed chemically pure,
does not insulate sufficiently well for experiment.   I placed a
sample purchased from Hopkin and Williams as pure, over quick-
lime, and then tested it.   At first it insulated fairly well, and
gave $K = 4\cdot75$.   In the course of a very few minutes $K = 4\cdot93$,
the insulation having declined so that observation was doubtful.
After the lapse of a few minutes more observations became impossible.   Professor Quincke in his first paper gives 4·623 and
4·660, and 4·394 in his second paper.

---

* *La Lumière Électrique*, vol. XXI. 1886, p. 97.
† *Wiedemann's Annalen*, vol. XXVIII. p. 474.

Bisulphide of Carbon. The sample was purchased from Hopkin and Williams, and tested as it was received—

$$K = 2\cdot67.$$

Professor Quincke finds $2\cdot669$ and $2\cdot743$ in his first paper, and $2\cdot623$ in his second. Palaz gives $2\cdot609$.

Amylene. Purchased from Burgoyne and Company—

$$K = 2\cdot05.$$

The refractive index (μ) for line D is $1\cdot3800$,

$$\mu^2 = 1\cdot9044.$$

Of the benzol series four were tested: *benzol, toluol, xylol,* obtained from Hopkin and Williams, *cymol* from Burgoyne and Company.

In the following table the first column gives my own results, the second those of Palaz, the third my own determinations of the refractive index for line D at a temperature of $17\cdot5°$ C., and the fourth the square of the refractive index :—

			μ.	μ^2.
Benzol	$2\cdot38$	$2\cdot338$	$1\cdot5038$	$2\cdot2614$
Toluol	$2\cdot42$	$2\cdot365$	$1\cdot4990$	$2\cdot2470$
Xylol	$2\cdot39$	—	$1\cdot4913$	$2\cdot2238$
Cymol	$2\cdot25$	—	$1\cdot4918$	$2\cdot2254$

For benzol Silow found $2\cdot25$, and Quincke finds $2\cdot374$.

The method employed by Palaz is very similar to that employed by myself in these experiments; but, so far as I can ascertain from his paper, he fails to take account of the induction between the case of his fluid condenser and his connecting wire; he also supports the inner coating of his fluid condenser on ebonite; and, so far as I can discover, fails to take account of the fact that this also would have the effect of diminishing to a small extent the apparent specific inductive capacity of the fluid. Possibly this may explain why his results are in all cases lower than mine. Determinations have also been made by Negreano (*Comptes Rendus,* vol. 104, 1887, p. 423) by a method the same as that employed by myself.

Three substances have been tried with the guard-ring condenser —double extra dense flint-glass, paraffin wax, and rock-salt. The first two were not determined with any very great care, as they were only intended to test the convenience of the method. For double extra dense flint-glass a value 9·5 was found ; the value I found by my old method was 9·896. For paraffin wax 2·31 was obtained—my previous value being 2·29. In the case of rock-salt the sample was very rough, and too small; the result was a specific inductive capacity of about 18, a higher value than has yet been observed for any substance. It must, however, be received with great reserve, as the sample was very unfavourable, and I am not quite sure that conduction in the sample had not something to do with the result. In the experiments with the guard-ring condenser the disturbing effect of the connecting wire was not eliminated. My thanks are due to my pupil, Mr Wordingham, for his valued help in carrying out the experiments.

27.

ON THE CAPACITY AND RESIDUAL CHARGE OF DIELECTRICS AS AFFECTED BY TEMPERATURE AND TIME. By J. HOPKINSON, F.R.S. AND E. WILSON.

[From the *Philosophical Transactions of the Royal Society*, Series A, Vol. CLXXXIX., 1897, pp. 109—136.]

Received December 15, 1896,—*Read January* 28, 1897.

BEFORE describing the experiments* forming the principal subject of this communication, and their results, it may be convenient to shortly state the laws of residual charge.

Let x_t be the potential at any time t of a condenser, *e.g.*, a glass flask, let y_t be the time integral of current through the flask up to time t, or, in other words, let y_t be the electric displacement, including therein electric displacement due to ordinary conduction. If the potential be applied for a short time ω, let the displacement at time t, after time ω has elapsed from the application of force $x_{t-\omega}$, be $x_{t-\omega} \, \psi(\omega) \, d\omega$; this assumes that the effects produced are proportional to the forces producing them; that is, that we may add the effects of simultaneously-applied electromotive forces. Generalise this to the extent of assuming that we may add the effects of successively-applied electromotive forces, then

$$y_t = \int_0^\infty x_{t-\omega} \, \psi(\omega) \, d\omega.$$

This is nothing else than a slight generalisation of Ohm's Law, and of the law that the charge of a condenser is proportional to its

* These experiments were commenced in the summer of 1894, and we have to thank Messrs C. J. Evans and R. E. Shawcross for valuable assistance rendered during the period of their Demonstratorship in the Siemens Laboratory, King's College, London.

potential. Experiments were tried some years ago for the purpose of supporting this law of superposition as regards capacity. It was shown that the electrostatic capacity of light flint glass remained constant up to 5,000 volts per millimetre (*Phil. Trans.*, 1881, Part II., p. 365). The consequences of deviation from proportionality were considered (*Proc. Roy. Soc.*, 1886, vol. 41, p. 453; *supra*, p. 104), and it was shown that, if the law held, the capacity as determined by the method of attractions was equal to that determined by the method of condensers; this is known to be the case with one or two doubtful exceptions (*supra*, p. 111). Rough experiments have been made to show that residual charge is proportional to potential; they indicate that it is (*Phil. Trans.*, vol. 167, Part II.). The integral $y_t = \int_0^\infty x_{t-\omega}\,\psi(\omega)\,d\omega$ includes in itself ordinary conduction, residual charge and capacity. Suppose that from $t = 0$ to $t = t$, $x_t = X$, and before that time $x_t = 0$, then $y_t = X \int_0^t \psi(\omega)\,d\omega$, and $\dfrac{dy_t}{dt} = \psi(t)$; thus $\psi(t)$ is the conductivity after electrification for time t. It has of course been long known that in stating the conductivity or resistance of the dielectric of a cable, it is necessary to state the time during which it has been electrified; hence $\psi(t)$ is for many insulators not constant. $\psi(\infty)$ may perhaps be defined to be the true conductivity of the condenser, but at all events we have $\psi(t)$ as the expression of the reciprocal of a resistance measurable, if we please, in the reciprocal of ohms. For convenience we now separate $\psi(\infty) = \beta$ from $\psi(\omega)$ and write for $\psi(\omega)$, $\psi(\omega) + \beta$. If we were asked to define the capacity of our condenser we should probably say: "Suppose the condenser be charged to potential X for a considerable time and then be short-circuited, let Y be the total quantity of electricity which comes out of it, then Y/X is the capacity." If T be the time of charging $y_t = X \int_0^T \{\psi(\omega) + \beta\}\,d\omega$ at the moment of short circuiting; $y_t = X \int_t^{T+t} \{\psi(\omega) + \beta\}\,d\omega$ after time t of discharge. The amount which comes out of the condenser is the difference of these, or $Y = X \left\{ \int_0^T \{\psi(\omega) + \beta\}\,d\omega - \int_t^{T+t} \{\psi(\omega) + \beta\}\,d\omega \right\}$; if t be infinite $\psi(t) = 0$, and $Y = X \int_0^T \psi(\omega)\,d\omega$; or we now have capacity expressed as an integral of $\psi(\omega)$ and measurable in microfarads, and it appears that the capacity is a function of the time of charge increasing as the time increases. Experiments have been made for testing this point in the case of light flint glass, showing that the capacity was the same for 1/20000 second and for ordinary durations of time (*Phil. Trans.*, 1881, p. 356), doubtless because $\int_{1/20000}^{\infty} \psi(\omega)\,d\omega$ is small compared with $\int_{0}^{1/20000} \psi(\omega)\,d\omega$. Now

$$\int_{0}^{t} \psi(\omega)\,d\omega,$$

when t is indefinitely diminished, may be zero, have a finite value, or be infinite; in fact it has a finite value. The value of $\psi(\omega)$ when ω is extremely small can hardly be observed; but

$$\int_{0}^{t} \psi(\omega)\,d\omega,$$

when t is small, can be observed. It is therefore convenient to treat that part of the expression separately, even though we may conceive it to be quite continuous with the other parts of the expression. $\int_{0}^{t} \psi(\omega)\,d\omega$, when t is less than the shortest time at which we can make observations of $\psi(\omega)$, is the instantaneous capacity of the condenser. Call it K and suppose the form of ψ to be so modified that for all observed times it has the observed values, but so that $\int_{0}^{t} \psi(\omega)\,d\omega = 0$, when t is small enough.

Then $y_t = Kx_t + \int_{0}^{\infty} x_{t-\omega}\{\psi(\omega) + \beta\}\,d\omega$. Here the first term represents capacity, the second residual charge, the third conductivity, separated for convenience, though really all parts of a continuous magnitude. Suppose now our condenser be submitted to a periodically varying electromotive force, that

$$x_t = A \cos pt,$$

then

$$y_t = A \left\{ K \cos pt + \int_{0}^{\infty} \cos p\,(t - \omega)\,[\psi(\omega) + \beta]\,d\omega \right\}$$

$$= A \left\{ K \cos pt + \cos pt \int_{0}^{\infty} \cos p\omega\psi(\omega)\,d\omega \right.$$

$$\left. + \sin pt \int^{\infty} \sin p\omega\psi(\omega)\,d\omega \right\}.$$

The effect of residual charge is to add to the capacity K the term $\int_0^\infty \cos p\omega \psi(\omega)\, d\omega$, whilst the term $\sin pt \int_0^\infty \sin p\omega \cdot \psi(\omega)\, d\omega$ will have the effect of conductivity as regards the phases of the currents into the flask. Thus the nature of the effect will depend upon the form of the function $\psi(\omega)$. An idea may be obtained by assuming a form for $\psi(\omega)$, say $\psi(\omega) = \dfrac{C}{t^m}$, where m is a proper fraction. This is a fair approximation to the truth. Then

$$\int_0^\infty \cos p\omega \psi(\omega)\, d\omega = \Gamma(1-m)\cos(1-m)\,\pi/2/p^{1-m},$$

$$\int_0^\infty \sin p\omega \psi(\omega)\, d\omega = \Gamma(1-m)\sin(1-m)\,\pi/2/p^{1-m}.$$

If m is near to unity, capacity is almost entirely affected; otherwise the effect is divided between the two, and dissipation of energy will occur. It is interesting to consider what sort of conductivity a good insulator such as light flint glass, according to this view of capacity, residual charge, and conduction, would have at ordinary temperatures if we could measure its conductivity after very short times of electrification; if, in fact, we could extend the practice used for telegraph cables and specify that the test of insulation should be made after the one hundred millionth of a second instead of after one minute, as is usual for cables. The capacity of light flint measured with alternating currents with a frequency of two millions a second is practically the same as when measured in the ordinary way; that is, its capacity will be 6·7. Its index of refraction is 1·57 or $\mu^2 = 2\cdot46$, or say, 2·5. We have then to account for 4·2 in a certain short time. The current is an alternating current, and we may assume as an approximation that it will be the residual charge which comes out in one-sixth of the period which produces this effect on the capacity; therefore

$$\int_0^{1/12 \times 10^6} \psi(\omega)\, d\omega = \frac{4\cdot2}{6\cdot7} \times \text{capacity of the flask as ordinarily measured.}$$

The capacity of a fairly thin flask may be taken to be 1/1000 microfarad to 2/1000 microfarad; hence we may take

$$\int_0^{1/12 \times 10^6} \psi(\omega)\, d\omega$$

to be 10^{-9} farad; if $\psi(\omega)$ were constant during this time its value must be $12 \times 10^6 \times 10^{-9} = \frac{1}{80}$ ohms^{-1} about. The value of $\psi(\omega)$

is far from constant, and hence the apparent resistance of that extraordinarily high insulator, a flint-glass flask, must be, for very short times, but still for times enormously large compared with the period of light waves, much less than 80 ohms.

[Added 11th March, 1897.—Somewhat similar considerations are applicable to conduction by metals. Maxwell pointed out that the transparency of gold was much greater than would be inferred from its conductivity measured in the ordinary way. To put the same thing another way—the conductivity of gold as inferred from its transparency is much less than as measured electrically with ordinary times. Or the conductivity of gold increases after the application of electromotive force. Suppose then we have a current in gold caused by an electromotive force which is increasing, the current will be less than it would be if the electromotive force were constant, by an amount approximately proportional to the rate of increase. If u be the current, ξ the electromotive force, $u = \alpha\xi - \beta\dot{\xi}$, where α is the conductivity as ordinarily measured. This gives us the equation of light transmission

$$\alpha\dot{\xi} - \beta\ddot{\xi} = \frac{d^2\xi}{dx^2},$$

assuming that we have no capacity in the gold.

Professor J. J. Thomson gives as a result of some experiments by Drude that the capacity of all metals is negative. This conclusion is just what we should expect, if we assume, as Maxwell has shown, that the conductivity of metals increases with the time during which the electromotive force is applied.]*

* The optical properties of metals may be expressed in the following manner on the principle enunciated in the text:

If f be the electric displacement in a metal, and X the electric force, then assuming only the generalized form of Ohm's law :

$$f = \int_0^\infty X_{t-\omega} \{\sigma + \psi(\omega)\}\, d\omega,$$

where σ is the conductivity as ordinarily measured, and $\psi(\infty) = 0$. Hence

$$\frac{df}{dt} = \sigma X + \int_0^\infty \left(\frac{dX}{dt}\right)_{t-\omega} . \psi(\omega) . d\omega.$$

If the disturbance be of period $\frac{2\pi}{p}$, we may write $X = X_0 e^{ipt}$, and the equation last written becomes

$$\frac{df}{dt} = \sigma X_0 e^{ipt} + \int_0^\infty X_0 e^{ip(t-\omega)} . \psi(\omega) . d\omega$$
$$= X \{\sigma + ip(C - iS)\}$$
$$= (\sigma + pS) X + C \frac{dX}{dt},$$

The experiments herein described are addressed to ascertaining the effect of temperature, first on residual charge as ordinarily known, second on capacity as ordinarily known, third to examining more closely how determinations of capacity are affected by residual charge, fourth to tracing the way in which the properties of insulators can continuously change to those of an electrolyte as ordinarily known. The bodies principally examined are soda-lime glass, as this substance exhibits interesting properties at a low temperature, and ice, as it is known that the capacity of ice for such times as one-tenth of a second is about 80, and for times of one-millionth of a second of the order of 3 or less.

Residual Charge as affected by Temperature.

Experiments on this subject have been made by one of us which showed that residual charge in glass increases with temperature up to a certain temperature, but that the results became then uncertain owing to the conductivity of the glass increasing. These experiments were made with an electrometer, the charge set free in the flask being measured by the rate of rise of potential on insulation. We now replace the electrometer by a delicate galvanometer and measure the current directly without sensible rise of potential.

where
$$S = \int_0^\infty \sin p\omega \, . \, \psi(\omega) \, . \, d\omega$$

and
$$C = \int_0^\infty \cos p\omega \, . \, \psi(\omega) \, . \, d\omega.$$

The metal therefore behaves as though it had conductivity $\sigma + pS$, and capacity $4\pi C$. Drude's experiments show that C is negative for most metals, in which there is at least nothing surprising; though it does not appear to follow rigidly from the fact of metallic transparency.

If the force be increasing at a constant rate, instead of being periodic, we have

$$\frac{df}{dt} = \sigma X + \frac{dX}{dt} \int_0^\infty \psi(\omega) \, . \, d\omega.$$

The transparency of thin metal sheets makes it probable that $\psi(\omega)$ is always negative. It follows that $\int_0^\infty \psi(\omega) \, . \, d\omega$, or the apparent capacity under a force increasing at a constant rate, is also negative. This is the meaning of the statement in the text. The capacity there referred to is not the same as the capacity determined by Drude's experiments. [Ed.]

Fig. 1 gives a diagram of connexions. The glass to be experimented upon is blown into a thin flask F, with thick glass in the neck to diminish the effect of charge creeping above the level of the acid, and is filled with sulphuric acid to the shoulder; it is then placed in sulphuric acid in a glass beaker, which forms the inner lining of a copper vessel consisting of two concentric tubes between which oil is placed. Thermometers, $T_1 T_2$, placed in

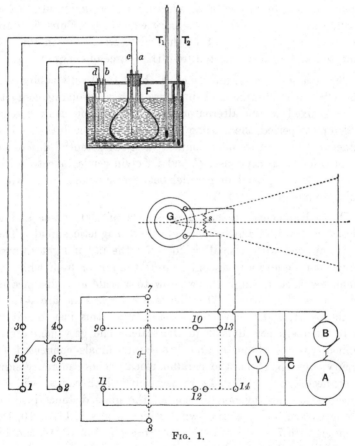

Fig. 1.

the acid outside the jar and in the oil, are made to register the same, or nearly the same, temperatures when taking observations, but T_1 gives the temperature taken for the flask. The flask is heated by a Bunsen burner placed under the copper vessel. Two electrodes a, c, insulated from one another and from the flask by means of sealing wax and glass tubes, dip into the sulphuric acid forming the inner coating of the jar, and similarly, electrodes b, d dipping into the outer acid make connexion with the outer coating. The acid inside and out was made to wet the flask up to a level higher than the acid would reach at the highest temperatures.

The four electrodes a, b, c, d, are connected respectively by thin copper wires, with four mercury cups 1, 2, 3, 4 cut in a block of paraffin, and, by means of a reversing switch, a, b and c, d can be connected respectively to mercury cups 5, 6. Cups 5, 6 are connected respectively to 7, 8 by thin wires, which can in turn be connected with or disconnected from the source of charge 9, 11.

The steady potential difference of about 1500 volts is obtained from a Siemens alternator A, in series with a revolving contact-maker B fixed to the alternator shaft and making contact once per complete period, there being six periods per revolution. The contact-maker is set to make contact when the potential difference is a maximum. A condenser C, and a Kelvin vertical electrostatic voltmeter V, are placed in parallel between the connecting wires leading to mercury cups 9, 11.

The galvanometer G has a resistance of 8000 ohms and is inclosed in an iron box which acts as a magnetic shield. The box is supplied with a small window for the ray of light to pass through it from an incandescent lamp to the mirror from which it is reflected back through the window to a scale at a distance of 12 feet from the mirror. The divisions on this scale are $\frac{1}{10}$th of an inch apart, and an average sensibility for this instrument is $\cdot 3 \times 10^{-9}$ ampere per division of the scale. The galvanometer is supplied with a shunt s, and has its terminals connected to mercury cups 13, 14 on the paraffin block. These mercury cups are connected to cups 10, 12 respectively, which can at will be connected to 7, 8, by one motion of the glass distance-piece g forming part of the reversing switch which places 9, 11, or 10, 12, in contact with 7, 8. A switch is so arranged that 13, 14 can be connected at will, that is, the galvanometer is short circuited.

The process of charging, discharging, and observing, is as follows :—Near the observer is a clock beating seconds which can be distinctly heard by the observer. Initially, the cups 9, 11, are disconnected from 7, 8; but 5, 1, and 6, 2, are connected. At the given moment the reversing switch is put over connecting 7, 9, and 11, 8; the jar is then being charged through electrodes a, b. This goes on for the desired time, during which charging volts and zero of the instrument are noted. At the end of the time required for charge, the main reversing switch is put over connecting 7, 10, and 8, 12; next the subsidiary switch is put over connecting 3 to 5 and 4 to 6, and on opening the short-circuiting switch, the spot of light is deflected and allowed to take up its natural state of movement determined by residual charge, readings being taken at stated epochs after discharge is started. This whole operation, including an adjustment of the shunt when necessary, was so speedily accomplished that reliable readings could be taken five seconds after discharge is started. By using two electrodes, polarization of electrodes is avoided, and the gradually-diminishing current through the galvanometer is that due to residual charge. The conductivity of the jar is determined by removing the glass distance-piece g, connecting 7 to 9, 8 to 12, and 10 to 11, and noting the steady deflection on the galvanometer for a given charging potential difference.

In the ice experiment, the conductors from 3, 4, are used both for charging and discharging. The form of condenser used when dealing with ice and liquid dielectrics is shown in Fig. 2. It consists of seven platinum plates, a, b, c, d, e, f, g, each measuring 2 inches by 3 inches, and of a thickness ·2 millim., separated from each other by a distance of 2·7 millims. To each plate are gold-soldered four platinum wires—two top and two bottom. Plates a, c, e, g, form the outer coating of the condenser, and are kept in their relative positions by cross connecting wires h, gold-soldered to the wires at each end of each plate. Similarly, plates b, d, f, which form the other and inner coating of the condenser, are fixed relatively to one another by cross connecting wires i. The relative positions of the two sets of plates are fixed by glass rods 1, 2. The terminals of the condenser are, for the inner plates the prolonged wire 3, and for the outer plates the wires 4, 4. These are bent round glass rods 5, 6, which resting on the top of a beaker support the plates in the fluid. The glass tubes on the wires 3, 4, 4, are for the purpose of securing good surface insulation. The glass beaker is conical, so as to remain unbroken when freezing the distilled water within. This was accomplished by surrounding the beaker with a freezing mixture of ice and salt, the lower temperature being obtained by further cooling in carbonic acid snow.

The same blue flask, which was the subject of the earlier experiments, was mounted as shown in Fig. 1, and the residual charge observed for various temperatures. This glass is composed of silica, soda and lime; the colour is due to oxide of cobalt in small quantity.

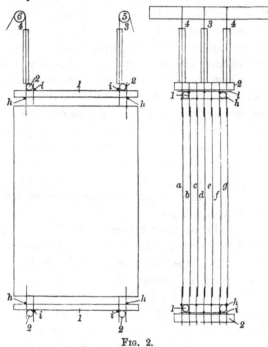

Fig. 2.

Out of a large number of experiments the data in Tables I. and II. give the general character of the results.

TABLE I.

Time in seconds	15° C.	34½	54½	70	85	117	132	Remarks
10	246	..	..	..	..	9770	7256	Blue flask. 6th and
15	..	376	1176	2785	5445	..	..	7th November,
20	121	265	1030	2586	4100	3590	3010	1894. Sensibility
30	87	209	892	2070	2980	2150	1735	of galvanometer,
60	46	131	683	1320	1510	950	778	·378 × 10⁻⁹. Du-
120	22½	91	483	720	688	440	350	ration of charge,
300	9¼	62	256	260	210	164	107	2 minutes. Charg-
							less than	ing volts, 1250.
600	..	..	123	110	..	86	59	

TABLE II.

Time in seconds	14° C.	55	70	110	137	Remarks
10	205	..	..	11740	12400	Blue flask. 13th November, 1894. Sensibility of galvanometer, ·407 × 10⁻⁹. Duration of charge, 2 minutes. Charging volts, 1250.
20	99	1230	2850	4790	4340	
60	38	837	1560	1212	990	
120	17	594	878	487	366	
300	5	308	314	134	..	

TABLE III.

	3rd January, 1895			4th January, 1895			Remarks
Time in seconds	8° C. 5 min. charge	117 1 min.	117½ 1 min.	8° C. 5 min. charge	122 1 min.	122½ 1 min.	
10	75	..	258	70	..	1100	New window-glass flask. Sensibility of galvanometer, ·358 × 10⁻⁹. Resistance of flask, 8° C., 3340 × 10⁶ ohms. Charging volts, 1500.
20	43	123	148	44	..	664	
30	30	105	109	32	400	..	
60	18	80	73	19	240	240	
180	7½	..	56	7¾	69	68	

The figures given are the deflections of the galvanometer in scale-divisions corrected for the shunt used. Recalling that one scale-division means a known value in ampères, that a known potential in volts is used, these figures can readily be reduced to ohms⁻¹. The capacity of the flask is 0·0026 microfarad at ordinary temperatures and times, and the specific inductive capacity of its material under similar conditions is about 8.   Hence one could reduce to absolute conductivities of the material. It is more interesting to consider how fast the capacity is changing. Take the first result given in Table III. for another flask 75 at 10 seconds; this means a conductivity 75 × 0·358 × 10⁻⁹/1500 = about 0·179 × 10⁻¹⁰, and this is, of course, the rate in farads per second at which the capacity is changing in that experiment compared with a capacity of the flask ½10⁻³ microfarad measured with the shortest times, or, to put it shortly, the flask owing to residual charge is changing capacity at the rate of about 3 per cent. per second. These figures also show that the residual charge up to

20 seconds increases greatly with the temperature; the residual at 60 seconds rises with the temperature up to about 70° C. or 80° C., and then diminishes; residual charge at 300 seconds begins to diminish at about 60° C. One may further note the way in which the form of the function $\psi(\omega)$ changes as temperature rises. Compare in Table I. the values for 20 and 30 seconds, the ratios are:—

Temperature	15	$34\frac{1}{2}$	$54\frac{1}{2}$	70	85	117	132,
Ratio		1·39	1·27	1·16	1·25	1·38	1·67	1·74.

In other words, if we expressed $\psi(\omega)$ in the form C/t^m, we should find m first diminishes as temperature rises to 54°, then increases as the temperature further rises. This has an important bearing upon the effect of residual charge on apparent capacity and resistance.

It will be noticed that the residual charge, for the same time, at high temperatures, is somewhat greater in Table II. than I. The results in Table I. were obtained on November 7th, 1894; those in Table II. on November 13th, 1894. There is no doubt but that heating this glass and submitting it to charge when heated, alters the character of the results in such manner as to increase residual charge for high temperatures. To test this more thoroughly, a new flask was blown out of window-glass composed of silica, lime and soda without colouring matter, and on January 3rd, 1895, was charged and discharged in the ordinary manner. After the results given in Table III. for January 3rd were obtained, the flask was charged for 21 minutes at 1500 volts, the direction of charge being reversed after 10 minutes, the temperature of the flask being 133°. We see that on January 4th, Table III., the same effect is observed, namely an apparent increase in residual charge for the same time at high temperatures. This may probably be attributed to a change in the composition of the material by electrolysis.

CAPACITY.

(A) *Low Frequency.*

Fig. 3 gives a diagram of connexions, showing how the apparatus is arranged for the purpose of determining the capacity of poor insulators, such as window-glass or ice, at varying temperatures. This is a bridge method, the flask F being placed in series with a condenser of known capacity K, and on the other side non-inductive resistances R_1, R_2. By means of keys k_1, k_2, the bridge can be connected to the poles of a Siemens alternator A; its potential difference is measured on a Kelvin multicellular voltmeter V. On the shaft of the alternator is fixed the revolving contact-maker B, which makes contact once in a period, and the epoch can be chosen.

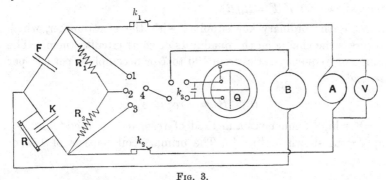

FIG. 3.

The Kelvin quadrant electrometer Q has one pair of quadrants connected to a pole of the revolving contact-maker B, and the other to a mercury cup 4 in a block of paraffin. The other terminal of B is connected to the junction between F and K; by means of mercury cups 1, 2, 3, the electrometer can be connected through the contact-maker to either end, or to the middle of the bridge.

The compensating resistance R is the resistance due to pencil lines drawn on a fine obscured glass strip*, about 12 inches long and $\frac{3}{4}$ inch wide, contact being made at each end by means of mercury in a small paraffin cup, and the whole varnished whilst hot with shellac varnish. A series of these resistances was made, ranging in value from a few megohms to a few tens of thousands of ohms. For the purpose of these experiments a knowledge of their actual resistance is of no moment, although for the purpose of manipulation their resistances are known.

The method of experiment is as follows:—Mercury cups 1 and 4 are connected by a wire, placing the electrometer and contact-

* See *Phil. Mag.* March, 1879.

maker across F, and the contact-maker is moved until it indicates no potential. Cups 3, 4 are now connected, and resistance R is adjusted until the electrometer again reads zero. After a few trials, alternately placing the bridge between 1, 4 and 3, 4, and adjusting R, the potentials are brought into the same phase, that is, the potential across the electrometer is zero in each case for the same position of the contact-maker. Mercury cups 2, 4 are now connected, the contact-maker B is adjusted to the point of maximum potential, and R_1, R_2 adjusted until balance is obtained. We now know that $K/F = R_1/R_2$.

k_3 is the ordinary key supplied with the electrometer, which reverses the charge on the quadrants or short circuits them. The range of frequency varies from 100 to 7 or 8 complete periods per second.

(B) *High Frequency.*

For high frequencies a method of resonance is used*, and the apparatus shown in Fig. 4. The primary coil consists of 1, 9, or

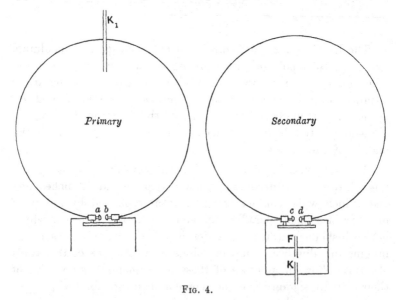

FIG. 4.

160 turns of copper wire 4 feet in diameter, having a condenser K_1 in its circuit and two adjustable sparking knobs a, b. The

---

* This method, we find, has been used by Thwing, *Physical Society's Abstracts*, vol. I. p. 79.

secondary is placed with its plane parallel to that of the primary, and usually at a distance of 4 or 5 feet from it; adjustable spark knobs c, d are provided in its circuit, which consists of 1, 9, or 160 turns of copper wire of the same diameter as the primary.   The diameters of the wires for the 1, 9, and 160 turns are respectively 5·3, 2·65, and 1·25 millims. A Ruhmkorff coil excites the primary. Between the spark knobs c, d are placed the capacity to be found F, and a large slide condenser K.   The method is one of substitution, that is to say, maximum resonance is obtained with both condensers attached by variation of K; F is removed and maximum resonance again obtained by increase of K.   In order to bring K on the scale for the two maxima, it is necessary to adjust K_1, the condenser in the primary.   This condenser consists of a sheet of ebonite with tin-foil on either side; three such condensers are available, and by variation of the area of tin-foil, if necessary, a suitable value for K_1 was speedily obtained.   Platinum-foil was used for the electrodes in the acid inside and outside the jar in the glass experiments instead of wire, as shown in Fig. 1, in order to secure that the connexions should not add materially to the self-induction of the circuit.

The frequency is calculated from the formula

$$\text{Frequency} = \frac{1}{2\pi \sqrt{KL 10^{-15}}},$$

where K is the capacity in secondary in microfarads,

L is the self-induction in centimetres.

$$L = 4\pi n^2 a \left(\log_e \frac{8a}{r} - 2 \right),$$

where n is the number of turns on the secondary,

$2a$ is the diameter of the ring = 4 feet,

$2r$ is the diameter of wire on secondary.

When $n = 1$, $L = 4230$ centimetres.   If K be taken ·00096 microfarad the frequency is $2·5 \times 10^6$.

The lowest frequency we have tried with this apparatus is when $n = 160$, $L = 136 \times 10^6$.   If K be taken ·0028 microfarad, the frequency is 8400.

That the capacity of some kinds of glass does not vary much with a moderate variation of temperature is known (*Phil. Trans.*, 1881, p. 365).   Experiments were tried on the same blue flask as

before, using the method in Fig. 3. The results obtained and
many times repeated for a frequency of 70 or 80 are given in
Table IV. As the specific inductive capacity of this flask,
measured in the ordinary manner, is about 8, it appears that
at 170° it is about 21. Knowing from the results in Tables I.
and II. how great was the residual charge for high temperatures
and short times, it appeared probable that the result would
depend upon the frequency. This was found to be the case,
as shown by the results of November 26, 1894, Table IV., the
apparent capacity being somewhat more than one-half at frequency
100 of what it is at frequency 7·3. Experiments on the window-
glass flask show the same result.

The next step was to determine whether or not this large
increase of apparent capacity was due to residual charge. To do
this the resonance experiments Fig. 4 were resorted to and the
capacity of the flask was determined with a frequency of about
2×10^6; it was found to be sensibly the same whether the flask
were hot or cold. The results show that the capacity varies from
185 to 198 in arbitrary units with a variation of temperature from
$25\frac{1}{4}°$ to 127°. With frequency 8400 the capacity varies from 240
to 285 in arbitrary units for a variation of temperature from 21°
to 122°, but here the sensibility was not so good as with the
higher frequency. We conclude that the apparently great capacity
of this glass at a temperature from 120° to 170° is due to residual
charge, but that the effects of this part of the residual charge are
not greatly felt if the frequency is greater than about 10,000
a second.

The capacity of window-glass is but little affected by variations
of frequency at ordinary temperatures.

TABLE IV.

20th November, 1894 Frequency, 72; volts, 70		21st November, 1894 Frequency, 85½; volts, 71½		26th November, 1894 Temperature, 120° C.		
Tempera- ture C.	Capacity of flask in terms of itself at 15° C.	Tempera- ture C.	Capacity of flask in terms of itself at 25° C.	Fre- quency	R_2/R_1	Remarks
15	1	25½	1	7·3	1·27	Standard conden-
92	1·31	54	1·05	12	1·11	ser unaltered
117	1·66	95	1·27	39½	·87	throughout ex-
154	2·6	120	1·59	71½	·78	periment
		170	2·61	100	·75	

CONDUCTIVITY AFTER ELECTRIFICATION FOR SHORT TIMES.

The Battery.—This consists of 12 series of small storage cells, Fig. 5, each series containing 50 cells. The poles of each set of 50 cells are connected to mercury cups in a paraffin block, and numbered 1, 3, 5, ... 21, 23, on the positive side; 2, 4, 6, ... 22, 24, on the negative. Cups b, d, are connected to the poles of the 56 cells in the Laboratory, and therefore, by connecting d, 1, 3, ... 21, 23, together on the one side, and 2, 4, ... 22, 24, b, together on the other side, the cells can be charged in parallel. For the purpose of these experiments, a large potential difference is required; this is obtained by removing the charging bars, and replacing them by a series of conductors connecting x to 1, 2 to 3 ... 22 to 23, 24 to y. In this manner, the whole of the 600 cells are placed in series with one another. Across the terminals x, y, are placed a condenser K_3 of about 4·3 microfarads, and a Kelvin vertical electrostatic voltmeter V. In order to change over quickly, and for the purpose of safety the charging bars and connexions for placing the cells in series are mounted on wood.

The Contact Apparatus.—This consists of a wooden pendulum carrying lead weights W_1, W_2, which were not moved during the experiments. The pendulum is released from the position p by the withdrawal of a brass plate, and, swinging forward, strikes a small steel contact piece f, carried on a pivoted arm of ebonite. The initial position of this ebonite arm is determined by a contact pin e, about $\frac{1}{16}$ inch diameter, contact being maintained by a spring m with an abutting rod insulated from a brass supporting tube by means of gutta-percha. This insulated rod is continued by a copper wire to the insulated pole of a quadrant electrometer Q. The brass supporting tube is continued by means of a metallic tape covering on the outside of the insulated wire, and is connected to the case and other quadrant of the electrometer. If, then, the pendulum be released from position p, the time which elapses between the terminal piece g first touching the plate f, and the time at which contact is broken between e and the insulated stop is the shortest time we have been able to employ in these experiments, its duration being ·00002 second.

For longer times an additional device, shown in plan only, is used. It consists of a brass pillar h, which carries a steel spring S, and which is moved to and fro in V-shaped slides by means of a screw provided with a milled head n, which is divided into twenty equal parts on the outside surface. A pointer fixed to the frame indicates the position of the head, and a scale on the brass slide shows the number of revolutions of the head from zero position. The pendulum steel piece g is of sufficient width to touch the spring S as it moves forward and strikes the plate f. The zero of the spring S is determined electrically by moving forward the pillar h, and noting the position of the milled head when contact is first made, the steel piece g being in contact with f, but not disturbing its initial position. The plate f is connected by a flexible wire with the slides which are in connexion with the spring S through its support h. When, therefore, the spring S leads the plate f by any distance, the time of contact is that time which elapses between g first striking S and the severance of contact between the pin e and its stop, always supposing that g keeps in contact with S. A good deal of trouble was experienced before making this contact device satisfactory. The ebonite arm carrying e and f was originally of metal, f being insulated; but inductive action rendered the results untrustworthy. Then again, the spring S, when first struck by the pendulum, evidently again severed contact before f was reached. To get over this difficulty a subsidiary series of fine steel wires were attached to S, so that as the pendulum moves forward the wires are one after the other struck. In order that the pendulum should not foul these wires or the spring S on its return to position p, it was slightly pressed forward by the hand at its central position.

The method adopted is that of the bridge. Starting from mercury cups x, we proceed by a fine wire to the terminal i, and thence, by a wire passing down the pendulum, to g. From g we pass through spring S and the piece f during contact to one end of the bridge. The flask F, or condenser to be experimented upon, is placed in series with metallic resistances a, these forming one arm of the bridge, the condensers K_1, K_2 forming the other arm. The stop e is connected to the junction between a and F; and the junction of K_1, K_2 is connected to the case of the electrometer by the outer conductor of the insulated wire leading to the instrument. The whole of the pendulum arrangement is supported on paraffin feet.

In the first instance pencil lines on glass were used for a, and K_1, K_2; but, for short times and varying current densities it was proved that these were unreliable, when a knowledge of their actual resistance at the time of contact is taken to be the same as measured in the ordinary way on a Wheatstone bridge.

Time of Contact.—The connexions were altered from those in Fig. 5 to those in Fig. 6. Eight dry cells having low internal

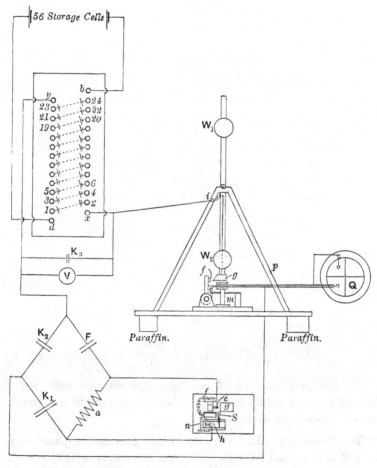

FIG. 5.

resistance were used for charging. In Fig. 6 let K be the capacity of the condenser, equal to $\frac{1}{3}$ microfarad. Let k be the capacity of the quadrant electrometer at rest in zero position, equal to ·000015 microfarad. Let R be the insulation resistance of K, and

r the resistance through which the condensers are charged. Let E be the E.M.F. of the battery, V be the E.M.F. of condenser, and t the time of contact in seconds.

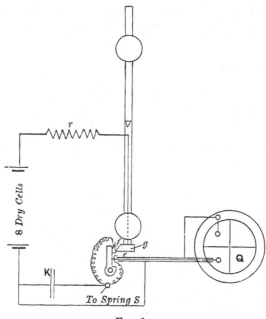

FIG. 6.

Then

$$(K + k) V + \frac{V}{R} = \frac{1}{r}(E - V),$$

$$V = \frac{RE}{R + r}\left\{1 - \exp.\left(-t \cdot \frac{R + r}{Rr} \cdot \frac{1}{K + k}\right)\right\}.$$

To determine E. Let $R = \infty$, $K = 0$, $r = 0$; the deflection of the electrometer needle from zero after the pendulum has struck gives E in scale-divisions.

To determine t. Let K be known and great as compared with k. Let $R = \infty$, and let r be such that the steady deflection from zero, V, after pendulum has struck, is about equal to half E.

$$V = E\left(1 - \epsilon^{-\frac{t}{Kr}}\right),$$

which gives

$$t = rK \log_\epsilon \frac{E}{E - V}.$$

The following are the values of t in seconds, so deduced, in terms of revolutions of the milled head n from zero :—

TURNS of milled head from zero

| 0 | $\frac{1}{2}$ | 1 | 2 | 3 | 4 | 5 | 6 |

TIME of contact in seconds

| ·00002 | ·00035 | ·00099 | ·0028 | ·006 | ·009 | ·011 | ·014 |

The experiments have so far dealt with frequencies ranging from 2×10^6 to 8000, and 100 to 10. The gap between 8000 and 100, during which the great effects of residual charge become apparent, is filled up by experiments with the pendulum apparatus just described. An attempt was made to fill up this gap by means of the method shown in Fig. 6, from which the effect on the capacity could be found for various times of contact, but this method was finally abandoned and used only for the determination of times of contact.

Referring to Fig. 5, F is the same window-glass flask mentioned above, and mounted as in Fig. 1; a is a non-inductive metal resistance, the effect of the capacity of which was at the most, when a is large, only capable of disturbing our experiments to the extent of eight per cent., but in most cases the disturbance is a small fraction of this; K_1 is a one-third microfarad condenser, and K_2 the large slide condenser used in the other experiments. The advantage of this method of experiment is that the charging potential difference V is great, and the actual ohmic resistance of a is small as compared with that of the flask F. In this manner the effect of the instantaneous capacity of the flask is overcome at once and the after effects due only to residual charge can be examined directly. The results are shown in Table V.

Let K_1, K_2, and F be discharged and let the potential difference V be applied to the bridge for time t. Let c be the ohmic resistance of the flask at the end of time t. Let K be its instantaneous capacity which is found by resonance at frequency 2×10^6. Let v be the potential across a. Then

$$\frac{v}{a} = \frac{V-v}{c} + K\frac{d}{dt}(V-v),$$

$$v = \frac{Va}{a+c}\left\{1 + \frac{c}{a}\,\epsilon^{-\frac{a+c}{acK}t}\right\}.$$

We know a, K, and t, and measure $\dfrac{a}{a+c}\left\{1+\dfrac{c}{a}\,\epsilon^{-\frac{a+c}{acK}t}\right\}$.

Now c is large compared to a, hence $\dfrac{a+c}{ac}=\dfrac{1}{a}$, therefore $\epsilon^{-\frac{a+c}{acK}t}$ is known; let it equal A. Then we have

$$\frac{a+Ac}{a+c}=\frac{K_2}{K_1+K_2}.$$

We have reduced a consistent with fair sensibility until the correction due to instantaneous capacity is so small as to be almost negligible, that is, until $\dfrac{c}{a}$ is sensibly equal to $\dfrac{K_1}{K_2}$.

How far we have been able to carry this can be seen by an inspection of Table V. It is only for the shortest time of contact that the correction for $\epsilon^{-\frac{t}{aK}}$ becomes at all sensible.

TABLE V.—*Window-Glass Flask.* 16th—31st October, 1896.

a.   Resistance box.

K_1. $\frac{1}{3}$rd $m.f.$ = 118,000 divisions of large slide condenser.

K_2. Large slide. When at zero = 100 of its own scale-divisions.

„   „   When at 435 = ·00146 $m.f.$

$K = $ ·0005 $m.f.$ from highest frequency resonance experiments.

In the diagram, Fig. 7, giving curves of conductivity and time for given temperature,

1 centim. vertical = 2×10^{-8} $(l^{-1}t)$.

1 centim. horizontal = 2×10^{-4} (t) seconds.

Therefore, area $\times\ 4 \times 10^{-6}$ gives capacity in microfarads.

TABLE V. (*continued*).—*Window-Glass Flask.* 16*th*—31*st October*, 1896.

Temperature of Flask, 15·4° C.

| Time of contact | | | | | Resistance of Flask c in 10^6 ohms | | $\frac{1}{c}$ in $10^{-6}(l^{-1}t)$ | $\int \frac{1}{c}dt - \frac{1}{c_\infty}t$ | | $F = K + \dfrac{\int \frac{1}{c}dt}{\frac{1}{c_\infty} - t}$ in m.f. |
Turns of milled head	Time in seconds	Large slide position K_2	a in 10^6 ohms	$\epsilon^{-\frac{t}{aK}} = A$	From $\frac{a+Ac}{a+c}$	From $a\frac{K_1}{K_2}$		Area in square centims.	m.f.	
0	·00002	160	·0056	·000794	3·99	2·5	·286			
0	·00002	430	·0090	·00117	2·80	2·0		11·0	·000044	·000544
·5	·00035			..	..	..		16·25	·000065	·000565
·5	·00099	160	·105	..	48	48	·0208	22·05	·000088	·000588
1·0	·00099									
1·0	·028	70	·130	..	90	90	·0111			
2·0	·0060									
3·0										
∞						1000 rough	·001			

54—57·3° C.

Turns of milled head	Time in seconds	Large slide position K_2	a in 10^6 ohms	$\epsilon^{-\frac{t}{aK}} = A$	From $\frac{a+Ac}{a+c}$	From $a\frac{K_1}{K_2}$	$\frac{1}{c}$ in $10^{-6}(l^{-1}t)$	Area in square centims.	m.f.	F
0	·00002	160	·00608	·00139	7·45	2·76	·200			
0	·00002	430	·0094	·00142	3·27	2·10	·085			
·5	·00035	430	·055	1/330000	12·2	12·2	·035	10·7	·000043	·000543
·5	·00035	160	·025	1/1·6 × 10¹²	11·4	11·4	·019	18·6	·000074	·000574
1·0	·00099	160	·065	..	29·5	29·5	·0059	26·2	·0001	·0006
1·0	·00099	160	·062	..	28·1	28·1				
2·0	·028	70	·075	..	52	52				
3·0	·0060					170				
∞										

TABLE V. (continued).—*Window-Glass Flask.* 16th—31st October, 1896.

Temperature of Flask, 80° C.

Turns of milled head	Time in seconds	Large slide position K_2	a in 10^6 ohms	$-\dfrac{t}{\epsilon\,aK}=A$	Resistance of Flask c in 10^6 ohms From $\dfrac{a+Ac}{a+c}$	From $a\dfrac{K_1}{K_2}$	$\dfrac{1}{c}$ in $10^{-6}(l^{-1}t)$	$\displaystyle\int\dfrac{1}{c}dt-\dfrac{1}{c_\infty}t$ Area in square centims.	m.f.	$F=K+\displaystyle\int\dfrac{1}{c}dt-\dfrac{1}{c_\infty}t$ in m.f.
0	·00002	160	·0041	·000058	2·52	1·86	·435			
0	·00002	430	·0079	·0063	2·04	1·80	·196	23·3	·000093	·000593
·5	·00035	430	·023	$1/16\times10^{12}$	5·1	5·1	·068	38·25	·000153	·000653
1·0	·00099	430	·066	:	14·7	14·7	·045	54·1	·000216	·000716
2·0	·0028	430	·100	:	22·3	22·3	·034	76·3	·000351	·000881
3·0	·0060	160	·065	:	29·5	29·5				
	8				65	65	·0154			

110—112° C.

Turns of milled head	Time in seconds	Large slide position K_2	a in 10^6 ohms	$-\dfrac{t}{\epsilon\,aK}=A$	Resistance of Flask c in 10^6 ohms From $\dfrac{a+Ac}{a+c}$	From $a\dfrac{K_1}{K_2}$	$\dfrac{1}{c}$ in $10^{-6}(l^{-1}t)$	$\displaystyle\int\dfrac{1}{c}dt-\dfrac{1}{c_\infty}t$ Area in square centims.	m.f.	$F=K+\displaystyle\int\dfrac{1}{c}dt-\dfrac{1}{c_\infty}t$ in m.f.
0	·00002	430	·0047	·00020	1·88	1·05	·714			
0	·00002	430	·0043	·000091	·973	·96	·256	24·5	·000098	·000598
·5	·00035	430	·0177	$1/16\times10^{16}$	3·9	3·9	·222	44·1	·000176	·000676
1·0	·00099	430	·020	:	4·5	4·5	·178	78·7	·000315	·000815
2·0	·0028	430	·025	:	5·6	5·6	·154	113·6	·000568	·00107
3·0	·0060	430	·029	:	6·5	6·5				
	8				8·5	8·5	·118			

TABLE V. (continued).—Window-Glass Flask. 2nd November, 1896.

Temperature of Flask, 126°C.

Time of contact										
0	·00002	430	·0035	1/89000	·779	·780	1·28	32·7	·000131	·000631
·5	·00035	430	·0082	..	1·83	1·83	·546	55·5	·000222	·000722
1	·00099	430	·0105	..	2·34	2·34	·427	70·8	·000283	·000783
2	·0028	430	·012	..	2·68	2·68	·373	102·2	·000511	·000101
3	·0060	430	·0133	..	2·97	2·97	·337	103	·000515	·00101
4	·009	430	·0135	..	3·0	3·0	·333	103	·000515	·00101
5	·011	430	·0135	..	3·0	3·0	·333			
∞						3·00	·333			

145°C.

Time of contact										
0	·00002	430	·0018	1/4·36 × 10⁹	·401	·401	2·49	50·8	·000255	·00075
·5	·00035	430	·0032	..	·713	·713	1·40	83	·00033	·00083
1	·00099	430	·004	..	·892	·892	1·12	93	·00037	·00087
2	·0028	430	·0043	..	·959	·959	1·04	93	·00037	·00087
3	·0060	430	·0043	..	·959	·959	1·04			
4	·009									
5	·011									

TABLE V. (*continued*).—10*th November*, 1896.

Window-Glass Flask. Solder used instead of Acid.

Time of con-tact	a in ohms	Resistance of flask c in 10^6 ohms	Temperature of flask °C.	$\frac{1}{c}$ in 10^{-6} ohms^{-1}
·00002	·7	·000156	about 350	6410
·00099	1·5	·000334	,, 350	3000
·0028	1·5	·000334	,, 350	3000
·011	1·5	·000334	,, 350	3000
·00002	18	·00446	285	224·0
·00002	130	·0290	229	34·5
·00002	200	·0446	219	22·4
·00002	270	·0602	203	16·6
·00099	330	·0736	202	13·6
·006	370	·0825	200	12·1
·014	370	·0825	200	12·1
Summary of re-sults with acid				
·00002	1000	·223	160	4·48
·00002	1800	·400	143	2·5
·00002	4700	1·05	112	·71
·00002	4100	1·86	80	·43
·00002	6080	2·76	55	·2
·00002	5600	3·99	15	·28

All temperatures from 15° to 145° were obtained by heating the flask as mounted in Fig. 1; for 200° to about 350° acid was taken away and a solder, melting at about 180° C., substituted. Since the solder only half filled the flask the conductivity should be about doubled for 200° to 350° when comparing with the lower temperatures.

Since $\frac{1}{c_t}$ is the conductivity of the jar at time t, let curves of conductivities be drawn in terms of contact in seconds. Fig. 7 gives these curves, which have been plotted from Table V. They show that, after a given time of contact, the effect of residual charge gradually diminishes as the temperature increases, until only the conductivity of the jar for infinite times is experienced. For instance, at about a temperature of 250° the table shows that the whole effect of residual charge has died away after 1/10,000 of a second. The total capacity of the jar at time t will be $K + \int_0^t \frac{1}{c}\,dt - \frac{1}{c_\infty}t$; where K is the instantaneous capacity which has been found by resonance to be $= \cdot0005$ microfarad for frequency 2×10^6.

$K_1 = 118{,}000$ divisions of the large slide condenser.

The curves in Fig. 7 have been integrated, and their area up to ·0028 second, when reduced to microfarads and added to K, shows that, for time of contact ·0028 second, the total capacity, which is ·000588 at temperature 15°·4, is ·00087 at temperature 145°. This total capacity diminishes as the times of contact

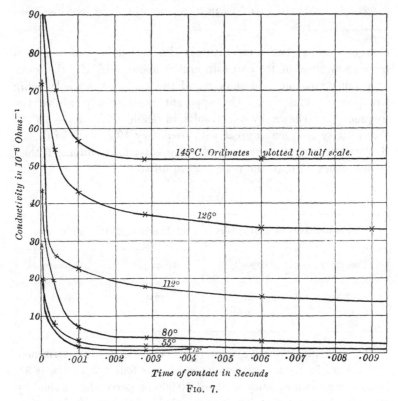

Fig. 7.

diminish, until we get to the results which resonance has shown; and then the capacity of this flask is sensibly the same for all temperatures when the frequency is of the order 2×10^6 per second.

ICE.

Ice was next examined, both in regard to its residual charge and its capacity. The residual charge is considerable, and increases as the temperature rises. Table VI. gives the residual

TABLE VI.

Time in seconds	About $-18°$ C.	About $-30°$ C.	Remarks
10	2800	866	Charging volts 890. 8th December, 1894
20	760	314	Duration of charge, $\frac{1}{4}$ minute in each case
60	377	74	Resistance at 945 volts
90	347	44	$-18°$ C., 7.2×10^6 ohms; $-30°$, 32.5×10^6 ohms

charge of ice at two temperatures: the higher is produced by a freezing mixture of ice and salt, and is about $-18°$ C.; the lower by placing carbonic acid snow round the beaker, the whole being wrapped in thick felt. The apparent capacity depends on the frequency, as shown by the results in Table VII. At $-18°$ C. the capacity is twice as great with frequency 10 as with 77·6. At the lower temperature the capacity is greater for frequency 9 than for frequency 77·6, in the ratio 1·39 to unity.

TABLE VII.

8th December, 1894, $-18°$ C. about		8th December, 1894, $-30°$ C. about	
Frequency	Capacity	Frequency	Capacity
77·6	·01	77·6	·0072
10	·019	9	·01

The specific inductive capacity of ice was next determined, with a high frequency, by resonance: it was found to be about 3*. Decreasing the frequency to about 10,000 rendered the method by resonance less sensitive, but it is certain that the specific inductive capacity is, for this frequency, of the order 3 rather than 50. We

---

* Thwing finds 2·85 to 3·36; Blondlot 2; Perrot 2·04.

conclude that the great deviation of ice from Maxwell's law is due to residual charge, which comes out between frequencies 10,000 and 100.

Our next step was to determine the resistance c, as in the case of glass, by the method shown in Fig. 5. The platinum plates, Fig. 2, were used, and to observe the temperature of the ice a platinum wire of resistance 1·32 ohms at 0° C. was frozen in the ice and surrounded the condenser. Table VIII. gives the results. K the capacity as given by the resonance experiments with frequency 2×10^6 was ·00022 microfarad. Adding to this $\int_0^t \frac{1}{c} dt - \frac{1}{c_\infty} t$, we find that at time ·0028 the total capacity is ·0038 at $-30°$ C., whereas it is for the same time ·0065 at $-18°$ C. The curves of conductivity are given in Fig. 8, and show the same character of results as those in the case of glass, Fig. 7.

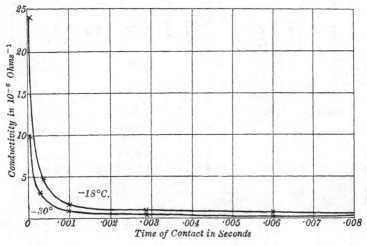

FIG. 8.

TABLE VIII.—*Ice.   5th November*, 1896.

a = Resistance box.

$K_1 = \frac{1}{3}$rd m.f.

K_2 = Large slide condenser.

K = Instantaneous capacity of ice condenser = ·00022 m.f.

TABLE VIII. (*continued*).—*Ice. 5th November, 1896.*

Temperature of Ice, $-18°$ C.

Time of contact		Large slide position K_2	a in 10^6 ohms	$\epsilon^{-\frac{t}{aK}} = A$	Resistance of Condenser c in 10^6 ohms		$\frac{1}{c}$ in 10^{-6}	$\int \frac{1}{c} dt - \frac{1}{c_\infty} t$		$K + \int \frac{1}{c} dt - \frac{1}{c_\infty} t$ in m.f.
Turns of milled head	Time in seconds				From $\frac{a+Ac}{a+c}$	From $a\frac{K_1}{K_2}$		Area in square centims.	m.f.	
0	·00002	430	·00019 / *·00022*	10^{-200}	·0424	·0424	23·6			
$\frac{1}{2}$	·00035	430	·00096 / *·0010*	...	·214	·214	4·67	16·8	·0034	·00362
1	·00099	430	·003 / *·003*	...	·669	·669	1·49	25·15	·005	·00525
2	·0028	430	·0059 / *·0060*	...	1·32	1·32	·758	31·4	·00628	·00650
3	·006	430	·0078 / *·0079*	...	1·74	1·74	·575	37·4	·0075	·00772
4	·009	430	·0095	...	2·12	2·12	·472	39·2	·0078	·00802
5	·011	430	·011 / *·011*	...	2·45 / *2·45*	2·45	·408	40	·008	·00822
6	·014	..	·011 / *·012*	...	2·45	2·45	·408	40	·008	·00822
7	..	..	*·013*	...	2·68	2·68	*·373*	40	·008	·00822
9	..	*430*	*·013*	...	2·90 / *2·90*	2·90	·345 / *·345*	40	·008	·00822

The italics give the result of a second experiment.

TABLE VIII. (continued).—Ice. 5th November, 1896.

				$1/7 \times 10^{-86}$	·100	·100	10	Temperature of Ice, −32° C. to −27° C.		
0	·00002	430	·00045	·		·100	10			·00208
½	·00035	430	·0015	·	·334	·334	2·99	9·3	·00186	·00306
1	·00099	430	·005	·	1·12	1·12	·893	14·2	·00284	·00380
2	·0028	430	·014	·	3·12	3·12	·320	17·9	·00358	·0047
3	·006	430	·021	·	4·68	4·68	·214	22·4	·0045	·0050
4	·009	430	·0255	·	5·69	5·69	·176	23·9	·0048	
5	·011	430								·0052
6	·014	·	·032	·	7·13	7·13	·140	25·1	·0050	
8	·	430	·039	·	8·7	8·7	·115			
10	·	·	·043	·	9·59	9·59	·104			
11	·	430	·045	·	10·0	10·0	·100			

[Added January 18th, 1897.]

CASTOR OIL.

This oil was obtained from Messrs Hopkin and Williams, and was tested as supplied. The platinum plates, Fig. 2, were submerged in this oil. Resonance experiments give, for frequency 2×10^6, a capacity equal to 105 divisions on the large slide condenser. For long times the method was not that shown in Fig. 3, but a bridge method, used in the earlier experiments[*], in which a Ruhmkorff coil is used for exciting. This test gives 139 scale divisions on the same slide condenser. In air the plates have capacity 30 scale divisions. We see, therefore, that at frequency 2×10^6 the specific inductive capacity would be 3·5 as against 4·63 for long times.

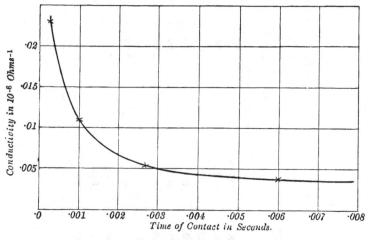

FIG. 9.

The short-time contact experiments, Fig. 5, give the results in Table IX., the temperature of the oil being 6° C., from which we see that residual charge in this oil is considerable. The total capacity after time of contact ·006 second is ·00034; whereas, with high frequency by resonance, it is ·000287 microfarad. The curve in Fig. 9 gives the relation between conductivity and time of contact, and has been plotted from Table IX.

---

[*] See *Proc. Roy. Soc.* vol. XLIII. p. 156. *Supra*, p. 112.

TABLE IX.—*Castor Oil.* 16*th November,* 1896.

$K_1 = \frac{1}{3}$ microfarad condenser; K_2 = large slide condenser.

$K = \cdot000287$ microfarad from High Frequency Resonance Experiment.

Temperature of Castor Oil, 6°C.

Time of contact		Large slide position K_2	a in 10^6 ohms	$\epsilon^{-\frac{t}{aK}} = A$	Resistance of Flask c in 10^6 ohms		$\frac{1}{c}$ in $10^{-6}(l^{-1}t)$	$\int \frac{1}{c}dt - \frac{1}{c_\infty}t$		$F = \dfrac{K + \int \frac{1}{c}dt}{1 - \frac{t}{c_\infty}}$ in m.f.
Turns of milled head	Time in seconds				From $\frac{a+Ac}{a+c}$	From $a\frac{K_1}{K_2}$		Area in square centims.	m.f.	
0	·00002	430	·015	1/100	− 2·7	3·34	·3*			
⅓	·00035	160	·095	1/361000	43·1	43·1	·023	6·2	·000025	·00031
1	·00099	40	·110	....	....	92·7	·0108	8·5	·000034	·00032
2	·0028	40	·223	....	....	188	·0053	12	·000048	·00033
3	·006	40	·320	....	....	270	·0037	14	·000056	·00034
5	·01	40	·460			388	·00258			
7	..	40	·620			523	·0019			
9	..	40	slightly greater than ·620							

* Taken from $a\frac{K_1}{K_2}$, since the negative value obtained from $\frac{a+Ac}{a+c}$ is untrustworthy, probably owing to K being still too large. To satisfy Maxwell's law, K should $= \cdot000176$ microfarad.

GLYCERINE.

This glycerine was obtained from Messrs Hopkin and Williams, and has been tested for purity and dried very carefully by Mr Herbert Jackson, of the Chemical Department of King's College, London.   The platinum plates, after careful cleaning in benzene, caustic potash, and water, were thoroughly dried and submerged in the glycerine in a beaker, the whole being placed in a glass receiver over a strong dehydrating agent.   After exhaustion, just sufficient air was admitted to render the space inside sufficiently non-conducting to stop discharge between the terminals of the condenser which are sealed into glass tubes supported by an india-rubber stopper.   The short-contact experiments show that the apparent resistance is 60,000 ohms, whether the time of contact be ·00002 or ·001 second, showing that there is no residual charge. The resonance experiments with high frequency give ·005 micro-farad for the capacity with glycerine, whereas with air the con-denser had ·000082 capacity; the specific inductive capacity is, therefore, about 60.   A test made as with castor oil with a Ruhmkorff coil at low frequency was difficult, but a fair approxi-mation was made by introducing a suitable compensating leakage into one of the other condensers of the bridge*.   The result indicated a capacity between 50 and 60.

WATER.

The platinum plates (Fig. 2) were placed in ordinary distilled water in a beaker which was cooled to 0° C. by a surrounding brine solution composed of water, common salt and ice.   The experiments with the short-contact apparatus show no material difference in the apparent resistance, whether the time of contact be ·00002 or ·00099 second; the apparent resistance for these times is 379 ohms.   The effects of residual charge in water do not affect the resistance within the range of times of contact given by this apparatus.

* This appears to have been done by Nernst, *Physical Society's Abstracts*, vol. I. p. 38.

[Added March 17th, 1897.]

OIL OF LAVENDER.

This oil was supplied by Messrs Hopkin and Williams: it was tested with the short-contact apparatus, Fig. 5, $K_1 = \cdot 33$, $K_2 = \cdot 0015$ microfarad. The charging potential was 1250 volts; the following figures give the results:—

Time of contact in seconds ·00002 ·00099   ·0028   ·006   ·01

a in ohms   .   .   . 9500 14000 14500 14800 14800

The high frequency resonance experiments give specific capacity 3·89: the frequency being of the order 2×10^6.

Two experiments were made at low frequency. First, the bridge method, Fig. 3, which gives the following results, the temperature of the oil being 16° C.:—

Frequency	Charging Volts	Specific Capacity
18	65	5·6
79	30	4·34

Second, the bridge method with a Ruhmkorff coil as used in the castor oil experiments. Temperature 14° C. Specific inductive capacity 4·18.

Experiments have been made by Stankewitsch (*Wied. Ann.*, 52), showing a variable capacity for oil of lavender. We, however, have not succeeded in obtaining any result so high as his.

28.

MAGNETISATION OF IRON.

[From the *Philosophical Transactions of the Royal Society*,
Part II., 1885, pp. 455—469.]

Received March 30,—Read April 23, 1885.

Preliminary.

THE experimental determination of the relation between
magnetisation and magnetising force would be a simple matter
if the expression of such relation were not complicated by the
fact that the magnetisation depends not alone on the magnetising
force at the instant, but also upon previous magnetising forces;
in fact, if it were not complicated by the phenomena of residual
magnetism. In the absence of any satisfactory theory, we can
only experimentally attack particular cases, and the results
obtained have only a limited application; for example, we may
secure that the sample examined has never been submitted to
greater magnetising force than that then operating, and we may
determine a curve showing the relation of magnetisation to
magnetising force when the latter is always increasing; we may
also determine the residual magnetism when after each experiment
the magnetising force has been removed. Such curves have been
determined by Rowland (*Phil. Mag.*, Aug., 1873) and others. For
many purposes a more useful curve is one expressing the relation
of the magnetising force and magnetisation when the former is
first raised to a maximum and then let down to a defined point;

ON THE MAGNETISATION OF IRON.		155

such curves have been called descending curves. One or two descending curves are given in a paper by Mr Shida (*Proc. R. S.*, 1883, p. 404). It has been shown by Sir W. Thomson and others that the magnetisation of iron depends greatly upon the mechanical force to which the iron is at the time submitted. In the following experiments the samples were not intentionally submitted to any externally applied force. Clerk Maxwell gives in his *Electricity and Magnetism*, chap. 6, vol. II., a modification of Weber's theory of induced magnetism, and from this he deduces, amongst other things, what had been already observed, that iron may be strongly magnetised and then completely demagnetised by a reversed force, but that it will not even then be in the condition of iron which has never been magnetised, but will be more easily affected by forces in one direction than in the other. This I have verified in several cases. The ordinary determinations of residual magnet- isation are not applicable to determine the permanent magnetism which a piece of the material of suitable given form will retain after removal of external magnetising force, but, as will be shown, the descending curves which express the relation of magnetisation and force, where these are diminishing, can be at once used for this purpose. Such curves can also be used, as has been shown by Warburg and by Ewing (*Report Brit. Assn.*, 1883), to determine the energy dissipated when the magnetisation of iron is reversed between given limits. That such dissipation must occur is clear, but some knowledge of its amount is important for some of the recent practical applications of electromagnetism. Probably Pro- fessor Ewing has made a more complete experimental study of magnetisation of iron than any one else. The researches of Professor Hughes should be mentioned here, as, although his results are not given in any absolute measure, his method of experiment is remarkable beyond all others for its beautiful simplicity. I have had great doubts whether it was desirable that I should publish my own experiments at all. My reason for deciding to offer them to the Royal Society is that a consider- able variety of samples have been examined, that in nearly all cases I am able to give the composition of the samples, that the samples are substantial rods forged or cast and not drawn into wire, and that determinations of specific electric resistance have been made on these rods which have some interest from a practical point of view.

Method of experiment.

Let $\mathfrak{H}$ be the magnetic force at any point, $\mathfrak{B}$ the magnetic induction, and $\mathfrak{I}$ the magnetisation (*vide* Thomson, reprint, Maxwell, vol. II., *Electricity and Magnetism*), then $\mathfrak{B} = \mathfrak{H} + 4\pi\mathfrak{I}$. We may therefore express any results obtained as a relation between any two of these three vectors; the most natural to select are the induction and the magnetic force, as it is these which are directly observed. $\mathfrak{B}$ is subject to the solenoidal condition, and consequently it is often possible to infer approximately its value at all points, from a knowledge of its value at one, by guessing the form of the tubes of induction. $\mathfrak{H}$ is a force having a potential, and its line integral around any closed curve must be zero if no electric currents pass through such closed curve, but is equal to $4\pi c$ if c be the total current passing through the closed curve. In arranging the apparatus for my experiments, I had other objects in view than attaining to a very small probable error in individual results. I wished to apply with ordinary means very considerable magnetising forces; also to use samples in a form easily obtained; but above all to be able to measure not only changes of induction but the actual induction at any time. The general arrangement of the experiments is shown in Fig. 1, and the apparatus in which the samples are placed in Fig. 2. In the latter fig. AA is a block of annealed wrought iron 457 millims. long, 165 wide, and 51 deep. A rectangular space is cut out for the magnetising coils BB. The test samples consist of two bars CC', 12·65 millims. in diameter; these are carefully turned, and slide in holes bored in the block, an accurate but loose fit; the ends which come in contact are faced true and square; a space is left between the magnetising coils BB for the exploring coil D, which is wound upon an ivory bobbin, through the eye of which one of the rods to be tested passes. The coil D is connected to the ballistic galvanometer, and is pulled upwards by an india-rubber spring, so that when the rod C is suddenly pulled back it leaps entirely out of the field. Each of the magnetising coils B is wound with twelve layers of wire, 1·13 mm. in diameter, the first four layers being separate from the outer eight, the two outer sets of eight layers are coupled parallel, and the two inner sets of four layers are in series with these and with each other. The magnetising current therefore divides

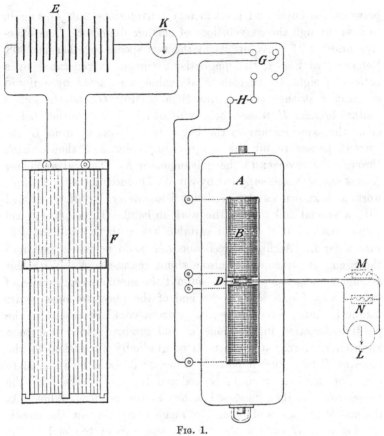

Fig. 1.

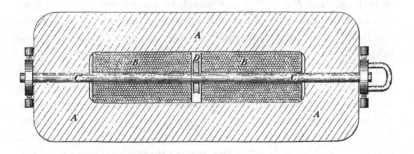

Fig. 2.

between the outer and less efficient convolutions, but joins again
to pass through the convolutions of smaller diameter. The effective number of convolutions in the two spools together is 2008.
Referring to Fig. 1, the magnetising current is generated by a
battery of eight Grove cells E, its value is adjusted by a liquid
rheostat F, it then passes through a reverser G, and through a
contact breaker H, where the circuit can be broken either before
or at the same instant as the bar C is withdrawn; from H the
current passes round the magnetising coils, and thence back
through the reverser to the galvanometer K. The galvanometer
K was one of those supplied by Sir W. Thomson for electric light
work, and known as the graded galvanometer, but it was fitted
with a special coil to suit the work in hand. The exploring coil
D was connected through a suitable key with the ballistic galvanometer L. Additional resistances M could be introduced into
the circuit at pleasure, and also a shunt resistance N. With this
arrangement it was possible to submit the sample to any series of
magnetising forces, and at the end of the series to measure its
magnetic state; for example, the current could be passed in the
positive direction in the coils B, and gradually increased to a
known maximum; it could then be gradually diminished by the
rheostat F to a known positive value, or it could be reduced to
zero; or, further, it could be reduced to zero, reversed by the
reverser G, and then increased to any known negative value. At
the end of the series of changes of magnetising current, the circuit
is broken at H (unless the current was zero at the end of the
series), and the bar C is simultaneously pulled outwards. Three
successive elongations of the galvanometer L are observed. From
the readings of the galvanometer K, the known number of convolutions of the coils B, and an assumed length for the sample bars,
the intensity of the magnetising force $\mathfrak{H}$ is calculated. The
exploring coil D had 350 convolutions. From its resistance,
together with that of the galvanometer with shunts, the sensibility
of the galvanometer, its time of oscillation, and its logarithmic
decrement, a constant is calculated which gives the intensity of
induction in the iron from the mean observed elongation of the
galvanometer. The resistances have been corrected in the calculation for the error of the B.A. unit, and both galvanometers were
standardised on the assumption that a certain Clark's cell had an
electromotive force of $1·434 \times 10^8$ C.G.S. units. This Clark's cell

had been compared to and found identical with those tested by
Lord Rayleigh.

Let the mean length of the lines of induction in the sample be
l, and σ the section of the sample; let l' be the length of lines of
induction in the block, and σ' their section, $\mathfrak{B}$ the intensity of
induction in the sample, $\mathfrak{B}'$ in the block, then $\sigma\mathfrak{B} = \sigma'\mathfrak{B}' = I$; let

$$\mathfrak{B} = \mu\mathfrak{H},$$

and

$$\mathfrak{B}' = \mu'\mathfrak{H};$$

then

$$\int\left(\frac{\mathfrak{B}}{\mu} + \frac{\mathfrak{B}'}{\mu'}\right) ds = \int\mathfrak{H}ds$$

or

$$\frac{Il}{\mu\sigma} = 4\pi nc - \frac{Il'}{\mu'\sigma'},$$

where n is the number of convolutions of the magnetising coils.
Now in the instrument used σ' is large, and μ' is as large as can
be obtained, hence the term $\dfrac{Il'}{\mu'\sigma'}$ is small comparatively.   My first
intention was to correct the magnetising force by deducting this
small correction, but finally I did not do so, because in the more
interesting results the magnetism of the block is dependent in
part upon previous magnetising forces, the effect of which cannot
be allowed for with certainty. We know then that in all the
curves the magnetising force indicated is actually too great by a
small but sensible amount, which does not affect the general
character of the results or their application to any practical
purpose. The magnetising force then at any point of the sample
is $\dfrac{I}{\sigma\mu} = \dfrac{4\pi nc}{l}$ — a small correction which we deliberately neglect.
There is another source of uncertainty in the magnetising force:
the length l is certainly greater than the space within the wrought
iron block, but it is not possible to say precisely how much greater.
If the sample bars and the block were a single piece, the results
of Lord Rayleigh for the resistance of a wire soldered into a block
would be fairly applicable; but it is essential that there should be
sufficient freedom for the bar to slide in the hole; the minute
difference between the diameters of the sample and the hole will
increase the value which should be assigned to l. Throughout, l

is assumed to be 32 centims., and it is not likely that this value is incorrect so much as half the radius of the bar, or 1 per cent. The magnetising forces ranged up to 240 C.G.S. units when both bars were of the same material. In some cases a single bar only was available for experiment; the plan then was to use it as the bar which enters into the exploring coil, and for the other to use a known bar of soft iron. We have then to deduct from $4\pi nc$ the magnetising force required to magnetise the bar of soft iron to the state observed, and to distribute the remainder over the shorter length of sample examined. The results obtained in this way are subject to a greater error, because some lines of induction undoubtedly make their way across from the end of the soft iron bar to the body of the block. A small correction is required, important in the case of bodies but slightly magnetic, for the fact that the area of the exploring coil is greater than the area of the bars tested. Thus the induction measured by the exploring coil is not only that in the sample, but something also in the air around the sample. The amount of this was tested by substituting for a sample of iron or steel a bar of copper, and afterwards a rod of wood, and it was found in both cases that the induction $\mathfrak{B}$ was 370 when the force $\mathfrak{H}$ was 230. The correction is in all cases small, but it has been applied in the column giving the maximum induction, as it materially affects the result when the sample contains much manganese, and is consequently very little magnetic.

The resistances were determined by the aid of a differential galvanometer. The resistances actually measured are, some of them, as low as $\frac{1}{6000}$ of an ohm, they must not therefore be regarded as so accurate as determinations made upon samples of a more favourable form; they, however, do show the remarkable effect of several impurities in iron, though it is possible that some of the results may be in error nearly 1 per cent.

Results obtained.

In all, thirty-five distinct samples were tested, of twenty compositions. The first three were supplied to me by Messrs Mather and Platt, and of these I have no analyses. All the rest were analysed for me in the laboratory of Sir Joseph Whitworth and Co., and the samples of material were actually prepared there, excepting Hadfield's steel, No. X.; Bessemer iron made by the basic process for telegraph wire, No. IV., from the North-Eastern Steel Company; and two Tungsten steels, Nos. XXX. and XXXI., which are in general use for permanent magnets.

I would express here my great indebtedness to Mr Gledhill, one of the managing directors of Sir J. Whitworth and Co., for preparing for me the samples I desired, and having them analysed. The fact is, indeed, that any value this paper may possess really lies in the variety of samples tried and in the accompanying chemical analysis, both due to Mr Gledhill. Samples Nos. I.—X. and XXXII.—XXXV. were tested with a pair of bars, the rest with a single bar of the sample used, in combination with a bar of wrought iron. The particulars of the several samples are most conveniently given in a table which follows, and to which I shall presently refer. With many samples observations were made sufficient to plot the ascending and descending curves which express induction in terms of magnetising force, but as these can make no pretence, for reasons already stated, to such accuracy as would warrant their use in testing a theory as to the form of curves of magnetisation, a few only are given as examples, and in other cases results are given in the table sufficient to define in absolute measure the primary magnetic properties of the materials and the very characteristic way in which they differ from each other.

The curves given include in each case an ascending curve, taken before the sample had been submitted to greater magnetising forces; a curve of residual magnetisation, that is, a curve in which the ordinate is the residual induction left after application and removal of the magnetising force represented by the abscissa, and two descending curves.

Fig. 3 gives the curves from wrought iron No. 1.

Fig. 4 the same to an amplified scale of abscissæ.

Fig. 5 for steel with ·89 per cent. carbon, annealed No. VIII.

Fig. 6 for steel with ·89 per cent. carbon, oil hardened No. IX.

Fig. 7 for cast iron No. III.

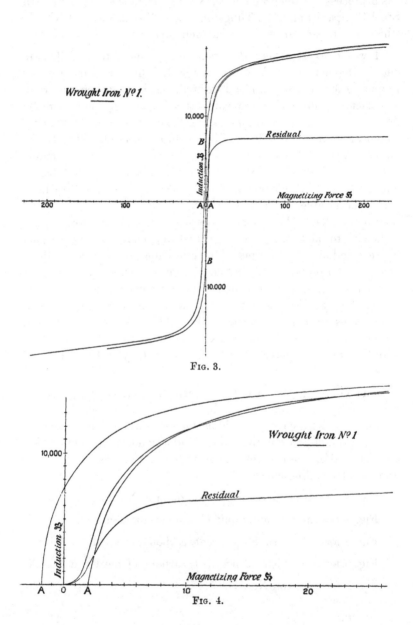

Fig. 3.

Fig. 4.

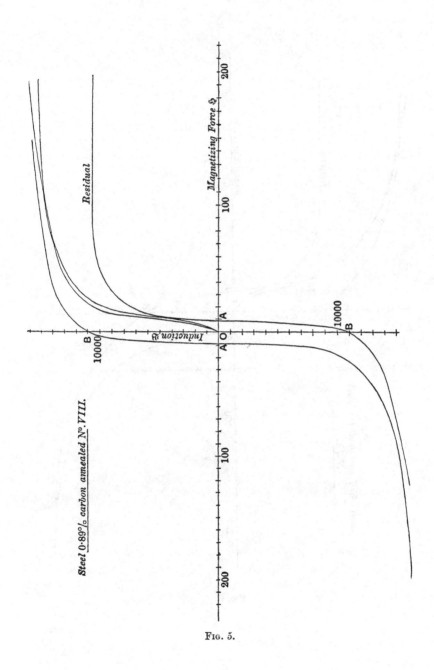

FIG. 5.

11—2

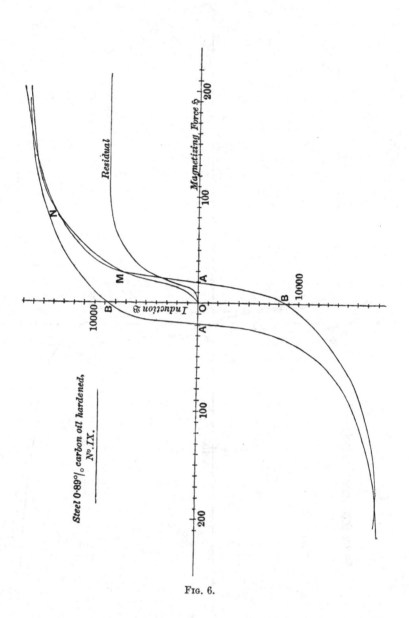

Fɪɢ. 6.

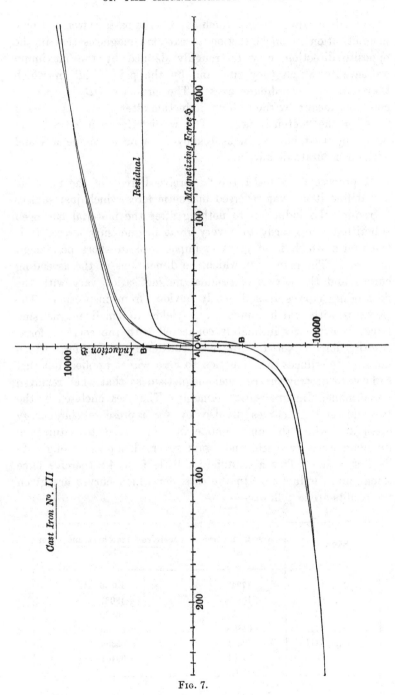

Fɪɢ. 7.

The descending curves, which express the passage from extreme magnetisation in one direction to extreme magnetisation in the opposite direction, may be roughly defined by the maximum ordinate to which they rise, and by the points AB in which they cut the coordinate axes. The ordinate OB is what is generally meant by the residual induction after great magnetising force, or the "retentiveness." The word "Coercive Force" has been long used, but, so far as I know, in a rather vague way and without accurate definition.

I propose to call OA the "Coercive Force" of the material, and define it as that reversed magnetic force which just suffices to reduce the induction to nothing after the material has been submitted temporarily to a very great magnetising force. It is the figure which is of greatest importance in short permanent magnets. The manner in which the dimensions of the ascending curves and the curves of residual magnetisation vary with the descending curves is sufficiently obvious from inspection. The slowness with which iron or steel yields to small magnetising forces is evidently intimately connected with the coercive force. Another force is worth noting, viz., that demagnetising force which not merely reduces the induction to zero whilst applied, but just suffices to destroy the residual magnetism so that when removed no permanent magnetisation remains. The area enclosed by the two descending curves divided by 4π represents the energy dissipated when the unit volume is magnetised to saturation, its magnetism reversed, and again reversed, and so brought to its first value. This area differs a little from $4 \times$ coercive force $\times$ maximum induction. In the cases for which curves are given the results are as follows :—

Sample	Area from curve $\dfrac{}{4\pi}$	4 coercive force × max. induction $\dfrac{}{4\pi}$
No. I.	17247	13356
,, III.	15139	13037
,, VI.	45903	40120
,, VII.	61898	65786
,, VIII.	50521	42366
,, IX.	74371	99401

In this we note that for soft iron the area is greater than the product, the reverse for hard steel; for any practical purpose we may assume that the greatest dissipation of energy which can be caused by a complete reversal to and fro of magnetisation is approximately measured by $\dfrac{\text{coercive force} \times \text{maximum induction}}{\pi}$.

An interesting feature in the curves is the manner in which the residual magnetism rapidly attains to near its maximum value, and is then nearly constant, whilst the induction continues to increase. This is very marked in the case of cast iron.

The column of figures in the general table of results almost explain themselves.

In the case of the cast iron, the total and the graphitic carbon are given, the difference being the combined carbon. In the case of the manganese steel and iron, the induction is almost proportional to the magnetising force, hence permeability is really the magnetic property to be noted: this is given below in a separate table. The demagnetising force is that reverse force which, when applied after great magnetising force, just suffices to remove all permanent magnetisation. The energy dissipated is

$$\frac{OA \times \text{maximum induction}}{\pi},$$

and is approximately the energy in ergs. converted into heat in a complete cycle of magnetisation from the limit in one direction to that in the opposite and back again.

In the general table of results one of the striking features is the high specific resistance of some samples of cast iron, ten times as great as wrought iron. This fact is not without practical importance in some forms of dynamo machines, for the energy wasted by local currents induced in the iron by given variations of the magnetic force will be but $\frac{1}{10}$th as great with cast iron as with wrought iron. The high resistance of cast iron may be due in large measure to its heterogeneity; grey cast iron may be regarded as a mechanical mixture of more or less pure iron with very small bits of graphite.

[Jan. 15, 1886.—I have recently determined the rate of variation with temperature of the electric resistance of a sample of cast iron for the purpose of ascertaining whether it approximated more

nearly to a pure metal, to an alloy, or to bodies the resistance of
which decreases with rise of temperature. The sample examined
was a thin rod of grey iron 6·71 millims. diameter and 24·85
centims. long between the contacts. The range of temperature
was 10° C. to 130° C., and through this range the rate of increase
of resistance was nearly uniform. The specific resistance at 0° C.
was inferred to be 0·000102, and the rate of increase was 0·00083
per degree centigrade.—J. H.]

Another very striking feature is the way in which any substantial proportion of manganese annihilates the magnetic property
of iron; the sample with 12 per cent. of manganese is practically
non-magnetic. The induction noted in the table = 310 corresponds
to a magnetising force of 244. If all the substances in this sample
other than the iron were mechanically mixed with the iron, and
arranged in such wise as to have the greatest effect upon its
magnetic property, no such annihilation of magnetic property
would ensue. This question of mixture will be considered somewhat more closely below.

The permeability and susceptibility are given in the following
table for some of the samples containing much manganese:—

No.	Permeability	Susceptibility
X.	1·27	·0215
XIV.	3·59	·206
XVI.	3·57	·2046
XXXV.	1·84	·0668

It is therefore clear that the small quantity of manganese
present enters into that which must be regarded for magnetic
purposes as the molecule of iron, and completely changes its
properties. The fact is one which must have great significance
in any theory as to what is the molecular nature of magnetisation.
Another clearly marked fact is the exceptionally great effect
which hardening has both upon the magnetic properties and the
electrical resistance of chrome steel.

Note also that in those cases where the maximum induction is
low, the residual magnetism is proportionately lower still, but that
the coercive force is not uniformly lower. This is in accordance

with the supposition that these samples are to be regarded as mechanical mixtures of a strongly magnetic substance, such as ordinary iron or steel, and a non-magnetic substance, such as manganese steel with 12 per cent. of manganese. A feature present in all the curves may some day have a bearing on the molecular theory of magnetism. It is this: the ascending curve twice crosses the continuation of the descending curve; in other words, the fact that a sample has been strongly magnetised in a reverse direction, renders it for small forces, or for large forces, more difficult to magnetise than a virgin sample, but distinctly easier for intermediate forces. This is best seen in the case of the hardest steel, No. IX., Fig. 6, in which the two curves cut in the points marked M, N. A similar phenomenon has been observed and investigated by G. Wiedemann (vide *Die Lehre vom Galvanismus*, first edition, vol. ii., p. 340, et seq.).

Magnetisation of a mixture of magnetic and non-magnetic substances.

We suppose that the mixture is purely mechanical, and that the two substances each retain their magnetic properties.

We may regard as an element of the substances a portion great in comparison with the size of the pieces of the two substances constituting the mixture, or we may be more analytical and regard as an element a portion very small in comparison with such pieces.

Let the volume of magnetic substance be λ, of non-magnetic $1 - \lambda$. The magnetic properties of the mixture will depend, not only upon λ, but upon the relative arrangement of the magnetic and non-magnetic parts.

Let α, a, A be the magnetic force, induction and magnetisation, regarding the sizes of the parts of the two substances as infinitely small; let α_0, a_0, A_0 be their values within a portion of magnetic substance. α, a, A are what we could actually observe. The relations of α_0, a_0, A_0 may be known from experiments on the magnetic substance when unmixed.

1. Suppose the magnetic substance to be arranged in the mixture in the form of filaments or laminæ parallel to the lines of

No.	Description	Temper	Chemical analysis					
			Total Carbon	Manganese	Sulphur	Silicon	Phosphorus	
I.	Wrought iron	Annealed	..	..	..	..	..	..
II.	Malleable cast iron	,,	..	..	..	..	..	Graphitic carbon 2·064
III.	Grey cast iron	,,	..	..	..	..	..	1·477
IV.	Bessemer mild steel	Annealed	·045	·200	·030	None	·040	..
V.	Whitworth mild steel	Annealed	·090	·153	·016	,,	·042	..
VI.	,, ,,	Oil hardened	·320	·438	·017	·042	·035	..
VII.	,, ,,	Annealed	,,	,,	,,	,,	,,	..
VIII.	,, ,,	Oil hardened	·890	·165	·005	·081	·019	..
IX.	,, ,,	Oil hardened	,,	,,	,,	,,	,,	..
X.	Hadfield's manganese steel	As forged	1·005	12·360	·038	·204	·070	..
XI.	Manganese steel	Annealed	·674	4·730	·023	·608	·078	..
XII.	,, ,,	Oil hardened	,,	,,	,,	,,	,,	..
XIII.	,, ,,	As forged	1·298	,,	·024	,,	·072	..
XIV.	,, ,,	Annealed	,,	8·740	,,	·094	,,	..
XV.	,, ,,	Oil hardened	,,	,,	,,	,,	,,	..
XVI.	,, ,,	As forged	·685	·694	,,	,,	,,	..
XVII.	Silicon steel	Annealed	,,	,,	,,	3·438	·133	..
XVIII.	,, ,,	Oil hardened	·532	,,	,,	,,	,,	..
XIX.	,, ,,	As forged	,,	·393	·020	·220	·041	..
XX.	Chrome Steel	Annealed	,,	,,	,,	,,	,,	Chromium ·621
XXI.	,, ,,	Oil hardened	·687	·028	,,	·134	·043	,,
XXII.	,, ,,	As forged	,,	,,	,,	,,	,,	,,
XXIII.	,, ,,	Annealed	,,	,,	,,	,,	,,	,,
XXIV.	,, ,,	Oil hardened	,,	,,	None	,,	,,	,,
XXV.	,, ,,	As forged	1·357	·036	,,	·043	·047	,,
XXVI.	Tungsten Steel	Annealed	,,	,,	,,	,,	,,	Tungsten 4·649
XXVII.	,, ,,	Hardened cold water	,,	·625	None	·021	·028	,,
XXVIII.	,, ,,	Hardened tepid water	·511	·312	,,	·151	·089	,,
XXIX.	,, ,,	Oil hardened	·855	,,	,,	,,	,,	,,
XXX.	,, ,, (French)	Very hard	,,	,,	,,	,,	,,	1·195
XXXI.	,, ,,		,,	,,	,,	,,	,,	,,
XXXII.	(Grey cast iron)		3·455	·173	·042	2·044	·151	Graphitic carbon 3·444
XXXIII.	Mottled cast iron		2·581	·610	·105	1·476	·435	2·353
XXXIV.	White cast iron		2·036	·386	·467	·764	·458	None
XXXV.	Spiegeleisen		4·510	7·970	Trace	·502	·121	..

No.	Description	Temper	Specific resistance	Maximum induction	OB, residual induction	OA, coercive force	Demagnetising force	Energy dissipated
					Magnetic properties *			
I.	Wrought iron	Annealed	·00001378	18,251	7,248	2·30	...	13,356
II.	Malleable cast iron	„	·00003254	12,408	7,179	8·80	...	34,742
III.	Grey cast iron	„	·00010560	10,783	3,928	3·80	...	13,037
IV.	Bessemer mild steel	„	·00001050	18,196	7,860	2·96	...	17,137
V.	Whitworth mild steel	Annealed	·00001080	19,840	7,080	1·63	...	10,289
VI.	„ „	„	·00001446	18,736	9,840	6·73	...	40,120
VII.	„ „	Oil hardened	·00001390	18,796	11,040	11·00	...	65,786
VIII.	„ „	Annealed	·00001559	16,120	10,740	8·26	...	42,366
IX.	„ „	Oil hardened	·00001695	16,120	8,736	19·38	...	99,401
X.	Halfield's manganese steel		·00005254	310				
XI.	Manganese steel	As forged	·00005368	4,623	2,202	23·50	37·13	34,567
XII.	„	Annealed	·00003928	10,578	5,848	33·86	46·10	113,963
XIII.	„	Oil hardened	·00005556	4,769	2,158	27·64	40·29	41,941
XIV.	„	As forged	·00006993	747				
XV.	„	Annealed	·00006316	1,985	540	24·50	50·39	15,474
XVI.	„	Oil hardened	·00007066	733				
XVII.	Silicon steel	As forged	·00006163	15,148	11,073	9·49	12·60	45,740
XVIII.	„	Annealed	·00006185	14,701	8,149	7·80	10·74	36,485
XIX.	„	Oil hardened	·00006195	14,696	8,084	12·75	17·14	59,619
XX.	Chrome steel	As forged	·00002016	15,778	9,318	12·24	13·87	61,439
XXI.	„	Annealed	·00001942	14,848	7,570	8·98	12·24	42,425
XXII.	„	Oil hardened	·00002708	13,960	8,595	38·15	48·45	169,455
XXIII.	„	As forged	·00001791	14,680	7,568	18·40	22·03	85,944
XXIV.	„	Annealed	·00001849	13,233	6,489	15·40	19·79	64,842
XXV.	„	Oil hardened	·00003035	12,868	7,891	40·80	56·70	167,050
XXVI.	Tungsten steel	As forged	·00002249	15,718	10,144	15·71	17·75	78,568
XXVII.	„	Annealed	·00002250	16,498	11,008	15·30	16·93	80,315
XXVIII.	„	Hardened cold water	·00002274	15,610	9,482	30·10	34·70	149,500
XXIX.	„	Hardened tepid water	·00002249	14,480	8,643	47·07	64·46	216,864
XXX.	„ (French)	Oil hardened	·00003604	12,133	6,818	51·20	70·69	197,660
XXXI.	„	Very hard	·00011400	9,148	3,161	13·67	17·03	39,789
XXXII.	Grey cast iron		·00005286	10,546	5,108	12·24		41,072
XXXIII.	Mottled cast iron		·00005661	9,342	5,554	12·24	24·40	36,383
XXXIV.	White cast iron		·00010520	385	77			
XXXV.	Spiegeleisen							

* The maximum magnetisation is obtained from the maximum induction by subtracting 240 (the magnetising force used) and dividing by 4π, or in most cases sufficiently nearly by multiplying by 0·08. The residual magnetism is obtained by dividing the residual induction by 4π. The numbers in the table are the actual magnetic forces in the infinitesimal gap between the two bars.

magnetic force, then $\alpha = \alpha_0$, and $A = \lambda A_0$. Hence the effect of admixture in this case is to reduce the magnetisation for a given force in the ratio $1 : \lambda$.

2. Let the non-magnetic substance be in thin laminæ lying perpendicular to the lines of force; we shall then have again $A = \lambda A_0$; but $a = a_0$ instead of $\alpha = \alpha_0$, whence

$$\alpha = a - 4\pi A$$
$$= a_0 - 4\pi \lambda A_0$$
$$= (1 - \lambda) a_0 + \lambda \alpha_0,$$

α_0 being supposed known in terms of a_0, this gives us the means of calculating the properties of the mixture.

These two are the extreme cases; all other arrangements of the two substances will have intermediate effects approximating to the one extreme or the other in a manner which we can judge in a rough way.

For example, if the magnetic substance be in separate portions bedded in the non-magnetic substance, the result will be somewhat analogous to the case of plates perpendicular to the lines of force; if on the other hand the non-magnetic substance be in separate portions bedded in the magnetic, the result will approximate rather to the case of filaments parallel to the lines of force.

Suppose that in the case of Hadfield's steel, No. X., the mixture be of pure iron in very small quantity in a non-magnetic matrix, how much pure iron is it necessary to suppose to be present, if the arrangement be as unfavourable as possible? Here $a = a_0 = 310$, $\alpha = 244$, $\alpha_0 =$ sensibly zero, whence $\lambda = \frac{66}{310} = 0\cdot21$. Suppose, however, that the iron were arranged as small spheres bedded in the non-magnetic substance, we have

$$\lambda = -\tfrac{1}{2} + \tfrac{3}{2} \, \frac{\mu}{2 + \mu},$$

μ being the observed value of $\dfrac{a}{\alpha}$, viz., $1\cdot27$; whence $\lambda = 0\cdot09$. We may say that of the 86 per cent. iron in this sample not more than 9 per cent. is magnetic.

If hard steel were bedded as small particles in a non-magnetic matrix, we should expect the mixture to have low retentiveness, but comparatively high coercive force, such as we see in the case of samples XI., XIII., and XV. If our apparatus had been sufficiently delicate to detect residual magnetism in samples X., XIV., and XVI., it is probable enough that we should have found the coercive force to be considerable.

In the case of mixtures much will depend on the relative fusibility of the magnetic and non-magnetic substances. If the former were less fusible, it would probably occur as crystals separated from each other by a non-magnetic matrix; if on the other hand it were more fusible, it would remain continuous. It is easy to see the kind of difference in magnetic property which would result.

Determination of permanent magnetisation of an ellipsoid.

If an ellipsoid be placed in a uniform magnetic field, its magnetisation will be uniform.

If the externally applied magnetising force be zero, the force at any point within the ellipsoid will be AL, BM, CN, where A, B, C are the components of magnetisation of the ellipsoid, and

$$L = 4\pi abc \frac{d\phi_0}{da^2}, \&c.,$$

where

$$\phi_0 = \int_0^\infty \frac{d(\phi^2)}{\sqrt{(a^2 + \phi^2)(b^2 + \phi^2)(c^2 + \phi^2)}},$$

and a, b, c are the semi-axes of the ellipsoid. Suppose the forces have all been parallel to the axis a, we have then $\mathfrak{H} = L\mathfrak{I} = \dfrac{L\mathfrak{B}}{4\pi}$ very nearly.

Let the curve PQ be the descending curve of magnetisation (the ordinates being induction), draw OR so that $\dfrac{RN}{ON} = \dfrac{4\pi}{L}$; then RN is clearly the induction in the ellipsoid when the external force is removed. In the case of a sphere $L = -\frac{4}{3}\pi$, therefore $RN = -3ON$. The greatest residual induction which a sphere of the materials can retain is a very little less than three times the force required to reduce the magnetisation to zero.

In a similar way any spheroid could be readily dealt with, and the best material judged for a permanent magnet of given propor-

tions. It should, however, be noted that any conclusions thus
deduced might be practically vitiated by the effect of mechanical
vibration in shaking out the magnetism from the magnet.

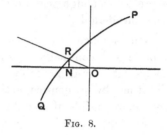

Dissipation of energy by residual magnetism.

Imagine a conducting circuit of resistance R, let x be the
current in it at time t, E the electromotive force other than that
due to the electro-magnetic field, and a the total magnetic induction through the circuit, then

$$Rx = E - \frac{da}{dt}.$$

The work done in time dt by the electromotive force is

$$xEdt = \left(Rx^2 + x\,\frac{da}{dt}\right)dt\,;$$

of this Rx^2dt goes to heat the wire, the remainder, or xda, goes
into the electro-magnetic field. Imagine a surface of which the
conducting circuit is a boundary, and on it take an elementary
area; through this area draw a tube of induction returning into

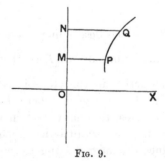

itself; the line integral of force along the closed tube is $4\pi x$. If
therefore we assume that the work done in any elementary volume

of the field is equal to that volume multiplied by the scalar of the product of the change of induction, and the magnetising force divided by 4π, the assumption will be consistent with the work we know is done by the electromotive force E. Now apply this to any curve connecting induction and magnetic force. Let PQ be two points in the curve, draw PM and QN parallel to the axis of magnetic force OX; the work done on the field per cubic centimetre passing from P to Q is equal to $\dfrac{\text{area } PMNQ}{4\pi}$. Some of this is converted into heat in the case of iron, for we cannot pass back from Q to P by diminishing the magnetising force.

Let AKB be the curve connecting A and B when the magnetising force is reversed, BLA when it is again reversed in this cycle; the final magnetisation is the same as it was initially;

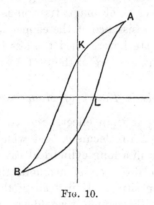

Fig. 10.

hence the balance of work done upon the field must be converted into heat; this heat will be represented by the area $AKBLA \div 4\pi$ in ergs. per cubic centimetre.

An approximation to the values of this dissipation is given in the table of results. It may be worth while to call attention to their practical application. Take the case of a dynamo-machine with an iron core, finely divided to avoid local electric currents. Note that we are going to assume—though whether true or false we do not know—that the dissipation is the same whether the magnetisation is reversed by diminishing and increasing the intensity of magnetisation without altering its direction, or whether it is reversed by turning round its direction without reducing its amount to zero.

A particular machine has in its core about 9000 cubic centims. of soft iron plates ; the resistance of its armature is 0·01 ohm, of its shunt magnets 8·0 ohms, and when running 900 revolutions per minute, its E.M.F. at the brushes is 55 volts. When the current in the armature is 250 ampères we have

Ergs. per second.

Total energy of current $= 144 \times 10^9$.

Loss in armature resistance $= 625 \times 10^7$.

Loss in magnet resistance $= 378 \times 10^7$.

Loss in magnetising and de-⎫ ⎧9000 cubic centims. × 15 revo-
   magnetising iron core of⎬ $=$ ⎨lutions per second × 13,356
   armature   ⎭ ⎩(from table of results) $= 18 \times 10^8$.

From this we see at once that the heat generated in the core of the armature by reversal of magnetisation is about one-half of that arising from the resistance of the copper wire of the electro-magnet. If a hard steel were used the loss from reversal might amount to 20 per cent. or more of the useful work done.

Weber's Theory of Magnetism.

In Weber's theory it is, in effect, assumed that the magnetic force tending to deflect a molecule is that which it would experience if it were placed in a long cylindrical cavity, the axis of the cylinder being in the direction of magnetisation. This seems a rather unnatural supposition. If instead of this we assume that the deflecting force is that which it would experience in a spherical cavity, and draw a curve connecting either the induction or magnetisation with the deflecting force on a molecule within a spherical cavity, we shall find that the curve differs very little from a straight line. In the curves already given we have taken $\mathfrak{B}$ and $\mathfrak{H}$ as the variables, where

$$\mathfrak{B} = \mathfrak{H} + 4\pi\mathfrak{J}.$$

Suppose we take $\mathfrak{B}$ and $\mathfrak{K}$ where

$$\mathfrak{B} = \mathfrak{H} + 4\pi\mathfrak{J},$$

and

$$\mathfrak{K} = \mathfrak{H} + \frac{4\pi}{3}\mathfrak{J}$$

$$= \tfrac{2}{3}\mathfrak{H} + \tfrac{1}{3}\mathfrak{B}.$$

The curves would then be hardly distinguishable from straight lines, the same scales being used for ordinates and abscissæ; it requires no great stretch of imagination to suppose that if this curve were continued far enough it would differ but little from that given by Maxwell, vol. ii., p. 79.

Now, in dealing with Weber's theory it would seem more suitable to take $\mathfrak{K}$, the magnetic force in a spherical cavity, as the independent variable. If we assume Weber's theory with this modification we arrive at the following conclusions:—

1. All observations yet made upon the magnetisation of iron are upon the straight part of Weber's curve.

2. The particular features of curves of magnetisation as ordinarily observed arise from a slight irregularity in Weber's curve, magnified by the near approach of iron to a state in which a random distribution of the magnetic axes of the molecules is unstable.

I do not put these remarks forward as indicating more than the fact that we are a very long way from obtaining a range of facts sufficiently extended for testing a molecular theory of magnetism. The broad fact which strikes the mind most forcibly is the specific difference which exists between magnetic and non-magnetic bodies. Most bodies are either very slightly ferro-magnetic or very slightly diamagnetic. On the other hand iron, nickel, and cobalt are enormously magnetic.

Iron with 12 per cent. of manganese, and some small quantities of carbon and other substances, is so little magnetic that its magnetism would be accounted for by supposing that in its mass were distributed a few little bits of pure iron. There seems to be a certain instability of something we know not what; bodies fall on one side practically non-magnetic, on the other enormously magnetic, but hardly any intermediate class exists.

The number of actual observations made on each of the samples named has been very considerable, though I have not thought it necessary to set them out at length, as I base no general conclusion upon them. The bulk of these observations were made by my assistant, Mr E. Talbot, and my pupil, Mr Paul Dimier, to whom my thanks are due for their patience and care.

29.

MAGNETIC PROPERTIES OF AN IMPURE NICKEL.

[From the *Proceedings of the Royal Society*, Vol. XLIV. pp. 317—319.]

Received June 9, 1888.

THE sample of nickel on which these experiments were made was supposed to be fairly pure when the experiments began. A subsequent analysis, however, showed its composition to be as follows:—

Nickel	95·15
Cobalt	0·90
Copper......................	1·52
Iron	1·05
Carbon.....................	1·17
Sulphur	0·08
Phosphorus	minute trace
Loss	0·13
	100·00

The experiments comprise determinations of the curve of magnetisation at various temperatures, the magnetising force being increased, that is to say, they are confined to a determination of the ascending curve of magnetisation. The temperature was always produced by enclosing the object to be tested in a double copper casing with an air space between the two shells of the casing, and by heating the casing from without by a Bunsen

burner.   The temperature was measured by determining the
electrical resistance of a coil of copper wire.   The copper was
first roughly tested to ascertain that its temperature coefficient
did not deviate far from ·00388 per degree centigrade of its
resistance at 20° C.; I was unable to detect that the coefficient
deviated from this value in either direction.   The temperature
may therefore be taken as approximately accurate.

The nickel had the form of a ring—Fig. 1.   On this ring were
wound in one layer 83 convolutions of No. 27 B.W.G. copper wire
carefully insulated with asbestos paper to serve as measurer of

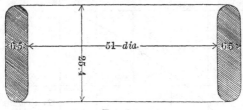

Fig. 1.

temperature and as secondary or exploring coil.   Over this again,
a layer of asbestos paper intervening, was wound a coil of 276
convolutions in five layers of No. 19 B.W.G. copper wire to serve
as the primary coil.

The method of experiment was simply to pass a known current
through the primary, to reverse the same and observe the kick
on a ballistic galvanometer due to the current induced in the
secondary.   At intervals the secondary was disconnected, and its
resistance was ascertained for a determination of temperature.
Knowing the current it is easy to calculate the magnetising force,
and knowing the constants of the galvanometer it is easy to
calculate the induction per square centimetre.   The practice was
to begin by heating the ring to a temperature at which it ceased
to be magnetic, then to lower the gas flame to a certain extent
and allow the apparatus to stand for some time, half-an-hour or
more, to allow the temperature to become steady, then determine
the temperature, then rapidly make a series of observations with
ascending force ; lastly, determine the temperature again.   The ring
was next demagnetised by a series of reversals with diminishing
currents.   The flame was further lowered, and a second series of
experiments was made.   It was then assumed that the previous

12—2

magnetisation would have a very small effect on any subsequent experiment. As the substance turned out to be far from pure nickel, it is not thought worth while to give actual readings. The results are given in the accompanying curves, Nos. I. to IV., in which the abscissæ represent the magnetising forces per linear

Curve I.

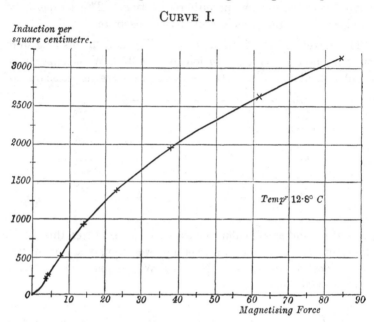

Curve II.

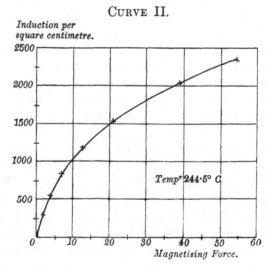

centimetre, the ordinates the induction per square centimetre, both in C.G.S. units. Curves V. and VI. give the results of Professor Rowland* for pure nickel at the two temperatures at which he

CURVE III.

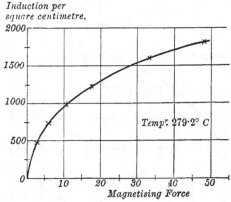

CURVE IV.

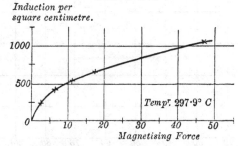

experimented. In Curves VII. and VIII. are given the inductions in terms of the temperature for stated intensities of the magnetising force, the ordinates being the inductions, the abscissæ the temperatures.

An inspection of these curves reveals the following facts :—

1. In my impure nickel much greater magnetising forces are required to produce the same induction than are required in Professor Rowland's pure nickel.

2. The portion of the curve which is concave upwards in my sample is less extensive and less marked than in his.

* *Phil. Mag.* November, 1874.

3. The magnetisation of my impure nickel disappears about 310° C.

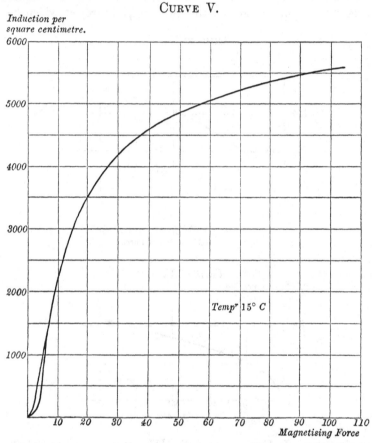

CURVE V.

4. A little below the temperature of 310° C. the induction diminishes very rapidly with increase of temperature.

5. At lower temperatures still the induction increases with rise of temperature for low forces, diminishes for high forces. This fact has been observed by several experimenters.

Specific Heat.—The object here was simply to ascertain whether or not there was marked change at the temperature when the nickel ceases to be magnetic. It appeared that this question could be best answered by the method of cooling, and that it mattered little even if it were roughly applied. A cylinder of nickel (Fig. 2) was taken, 5·08 cm. diameter, 5·08 cm. high, having

CURVE VI.

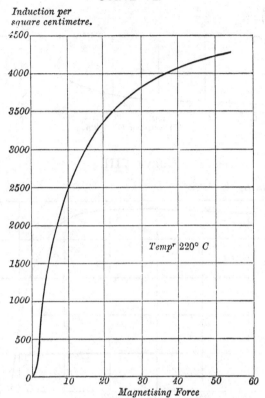

Induction per square centimetre.

Tempr 220° C

Magnetising Force

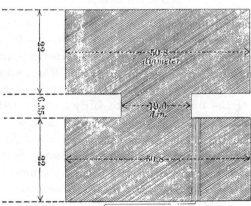

Hole for admission of copper wire.

FIG. 2.

CURVE VII.

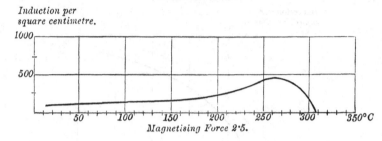

Induction per square centimetre.

Magnetising Force 2·5.

CURVE VIII.

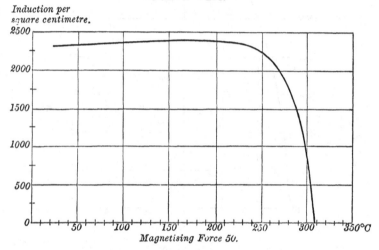

Induction per square centimetre.

Magnetising Force 50.

a circumferential groove, 15·9 mm. deep and 6·35 mm. wide. In this groove was wound a copper wire, well insulated with asbestos, by the resistance of which the temperature was determined. The cylinder was next enveloped in many folds of asbestos paper to insure that the cooling should be slow, and that consequently the temperature of the nickel should be fairly uniform and equal to that of the copper wire. The whole was now heated over a Bunsen lamp till the temperature was considerably above 310° C.; the lamp was next removed, and the times noted at which the resistance of the copper wire was balanced by successive values in the Wheatstone's bridge. If θ be the temperature, and t be time, and if the specific heat be assumed constant, and the rate of loss of heat proportional to the excess of temperature, $k\dfrac{d\theta}{dt} + \theta = 0$

or $k \log \theta + (t - t_0) = 0$. In Curve IX. the abscissæ represent the time in minutes, the ordinates the logarithms of the temperature, the points would lie in a straight line if the specific heat were constant. It will be observed that the curvature of the curve is small and regular, indicating that although the specific heat is not quite constant, or the rate of loss is not quite proportional to the excess of temperature, there is no sudden change at or about 310° C. Hence we may infer that in this sample there is no great or sudden absorption or liberation of heat occurring with the accession of the property of magnetisability.

CURVE IX.

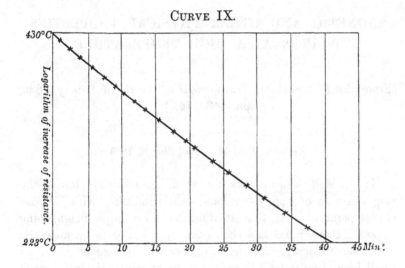

30.

MAGNETIC AND OTHER PHYSICAL PROPERTIES OF IRON AT A HIGH TEMPERATURE.

[From the *Philosophical Transactions of the Royal Society*, 1889, pp. 443—465.]

Received April 16,—*Read May* 9, 1889.

IT is well known that for small magnetising forces the magnetisation of iron, nickel, and cobalt increases with increase of temperature, but that it diminishes for large magnetising forces*. Bauer† has also shown that iron ceases to be magnetic somewhat suddenly, and that the increase of magnetisation for small forces continues to near the point at which the magnetism disappears. His experiments were made upon a bar which was heated in a furnace and then suspended within a magnetising coil and allowed to cool, the observations being made at intervals during cooling. This method is inconvenient for the calculation of the magnetising forces, and the temperature must have been far from uniform through the bar. In my own experiments‡ on an impure sample of nickel the curve of magnetisation is determined at temperatures just below the temperature at which the magnetism disappears, which we may appropriately call the critical temperature.

---

* Rowland, *Phil. Mag.* November, 1874.

† Wiedemann, *Annalen*, vol. XI. 1880.

‡ *Roy. Soc. Proc.* June, 1888. *Supra*, p. 178.

Auerbach* and Callendar† have shown that the electrical resistance of iron increases notably more rapidly than does that of other pure metals.   Barrett‡, in announcing his discovery of recalescence, remarked that the phenomenon probably occurred at the critical temperature.   Tait§ investigated the thermo-electric properties of iron, and found that a notable change occurred at a red heat, and thought it probable that this change occurred at the critical temperature.

It appeared to be very desirable to examine the behaviour of iron with regard to magnetism near the critical temperature, and to ascertain the critical temperatures for different samples.   It also appeared to be desirable to trace the resistance of iron wire up to and through the critical temperature, and to examine more particularly the phenomenon of recalescence, and determine the temperature at which it occurred.

The most interesting results at which I have arrived may be shortly stated as follows :—

For small magnetising forces the magnetisation of iron steadily increases with rise of temperature till it approaches the critical temperature, when it increases very rapidly, till the permeability in some cases attains a value of about 11,000.   The magnetisation then very suddenly almost entirely disappears.

The critical temperatures for various samples of iron and steel range from 690° C. to 870° C.

The temperature coefficient of electrical resistance is greater for iron than for other metals ; it increases greatly with increase of temperature till the temperature reaches the critical temperature, when it suddenly changes to a value more nearly approaching to other metals.   Recalescence does occur at the critical temperature.   The quantity of heat liberated in recalescence has been measured and is found to be quite comparable with the heat required to melt bodies.

Since making the experiments and writing the preliminary notes which have already appeared in the *Proceedings of the Royal Society*, my attention has been called to two papers

* Wiedemann, *Annalen*, vol. v. 1878.
† *Phil. Trans.* A, 1887.
‡ *Phil. Mag.* January, 1874.
§ *Edinburgh Roy. Soc. Trans.* December, 1873.

which deal in part with some of the matters on which I have been experimenting. Pionchon[*] has shown that the specific heat of iron is very much greater at a red heat than at ordinary temperatures. W. Kohlrausch[†], in an interesting paper, shows that, whereas the temperature coefficient of resistance of iron is much greater than usual for temperatures below the critical temperature, it suddenly diminishes on passing that temperature. He also identifies the temperature of recalescence with the critical temperature. So far as resistance of iron is concerned, W. Kohlrausch has anticipated my results, which I give, however, for the sake of completeness.

Magnetic Experiments.

The method of performing the magnetic experiments was the same as that used by Rowland. The copper wire was, however, insulated carefully with asbestos paper laid over the wire, and with layers of asbestos paper between the successive layers of the wire. The insulation resistance between the primary and the secondary coils was always tested, both at the ordinary temperature and at the maximum temperature used. At the ordinary temperature this resistance always exceeded a megohm; at the maximum temperature it exceeded 10,000 ohms, and generally lay between 10,000 and 20,000 ohms. The ring to be examined, with its coils of copper wire, was placed in a cylindrical cast-iron box, and this in a Fletcher gas furnace, the temperature of which was regulated by the supply of gas. The temperatures were estimated by the resistance of the secondary coil. It was observed that the resistance of this coil at the ordinary temperature increased slightly after being raised to a high temperature; this I attribute to oxidation of the wire where it leaves the cast-iron box. However, it introduced an element of uncertainty into the determination of the actual temperatures, amounting, perhaps, to 20° C. at the highest temperature. This error will not affect the differences between neighbouring temperatures, with which we are more particularly concerned.

[*] *Comptes Rendus*, vol. CIII. p. 1122.
[†] Wiedemann, *Annalen*, vol. XXXIII. 1888.

The resistance of the ballistic galvanometer is 0·43 ohm ; to this additional resistances were added to give the necessary degree of sensibility. The ratio of two successive elongations of the galvanometer is $(1 + r)/1 = 1·12/1$. The time of oscillation T and the sensibility varied a little during the experiments, but so little that the correction would fall within the limits of errors of observation in these experiments.

The total induction $= \left\{ \left(1 + \dfrac{r}{2}\right) \dfrac{C}{\alpha} \dfrac{T}{2\pi} \right\} \dfrac{1}{2n}\ RA\ 10^8$, where C is the current which gives the deflection α, n is the number of turns in the secondary coil, R the resistance of the secondary circuit, A the mean of the first and second elongations on reversal of the current in the primary.

The magnetising force $= 4\pi mc/l$, where m is the number of turns in the primary, l the mean length of lines of force in the ring, c the current in absolute measure in the primary.

With my galvanometer as adjusted, a Grove's cell, the E.M.F. of which was at the time determined to be 1·800 volt, gave a deflection of 158·5 divisions through a resistance of 50,170 ohms, whence

$$\frac{C}{\alpha} = \frac{1·800}{158·5 \times 50,170} = 0·0000002264,$$

$$T = 13·3.$$

Hence $\qquad \left(1 + \dfrac{r}{2}\right) \dfrac{C}{\alpha} \dfrac{T}{2\pi} = 5·09 \times 10^{-7}.$

The ring method of experiment is open to the objection that the magnetising force is less in the outer than in the inner portions of the ring. The results, in fact, give the average results of forces which vary between limits.

Wrought-Iron.—The sample of wrought-iron was supplied to me by Messrs Mather and Platt. I have no analysis of its composition. I asked for the softest iron they could supply*.

* [*Added July 2*, 1889.—Sir Joseph Whitworth and Co. have since kindly analysed this sample for me with the following result :—

	C	Mn	S	Si	P	Slag (containing 74 per cent. SiO_2)
Per cent.	·010	·143	·012	Nil	·271	·436.]

The dimensions of the ring were as shown in the accompanying sketch :—

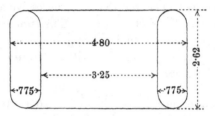

The area of section is 1·905 sq. cm. The area of the middle line of the secondary coil is estimated to be 2·58 sq. cms. This estimate is, of course, less accurate than the area of section of the ring itself.

The secondary coil had 48 convolutions, the primary 100 convolutions.

At the beginning of the experiments the insulation resistance of the secondary from the primary was in excess of 1 megohm ; the resistance of the secondary and the leads was 0·692, the temperature being 8°·3 C.

The resistance of the leads to the secondary and of the part of the secondary external to the furnace was estimated to be 0·04.

CURVE I.

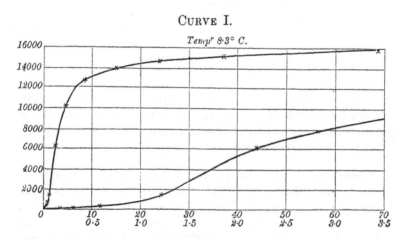

A curve of magnetisation was determined at the ordinary temperature on the virgin sample with the following results, shown graphically in Curve I.; in each case the observation was repeated twice with reversed direction of magnetising currents, and the kicks in the galvanometer were found to agree very closely together:

Magnetising force

0·15   0·3   0·6   1·2   2·2   4·4   8·2   14·7   24·7   37·2   69·2

Induction per sq. cm.

39·5   116   329   1,560   6,041   10,144   12,633   14,059   14,702   15,149   15,959

The ring was next heated and observations were made with a magnetising force of 8·0 to ascertain roughly the point at which the magnetism disappeared. After the magnetism had practically disappeared and the temperature was roughly constant, as indicated by the resistance, being 2·92 before the experiment and 2·85 after the experiment, corresponding with temperatures of 838° C. and 812° C., the induction was determined for varying magnetising forces.

Magnetising force	2·4	4·2	8·0	21·0	49·8
Total induction	small	12·3	22·7	58·2	143

This shows that the induction is, so far as the experiment goes, proportional to the inducing force.

Taking the total induction as 143, corresponding to a force of 49·8, we have induction in the iron 109, or 57 per sq. cm., giving permeability equal to 1·14, showing that the material has suddenly become non-magnetic.

The ring was now allowed to cool, some rough experiments being made during cooling. When cold the resistance of the secondary and the leads was found to be 0·697 ohm. The ring was again heated till the resistance of the secondary reached 2·845 and the magnetism had disappeared. It was next allowed to cool exceedingly slowly, and the following observations were made with a magnetising force of 0·075 C.G.S. unit:—

Resistance of secondary . .	2·81	2·80	2·79	2·78	2·765
Temperature	796°	792°	788°	785°	781°
Induction per sq. cm. . .	0	0	0	126·8	

showing that magnetisation returns at a temperature corresponding to resistance between 2·78 and 2·765.

Systematic observations then began.   The results are given in
the following tables and the curves to which reference is made.
The curves are in each case set out to two scales of abscissæ,
the better to bring out their peculiarities.

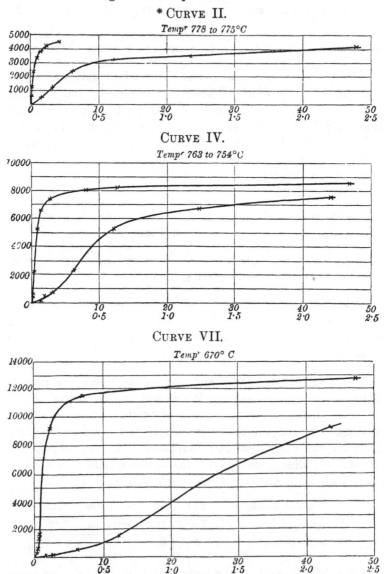

* CURVE II.

Tempr 778 to 775°C

CURVE IV.

Tempr 763 to 754°C

CURVE VII.

Tempr 670° C

* Though many of the tables and curves have been omitted, the enumeration of
those retained has been kept the same as in the original, to facilitate reference. [ED.]

TABLES 1, 3, 6, 8.

	Table 1, Curve II		Table 3, Curve IV		Table 6, Curve VII		Table 8, Curve IX	
Resistance of secondary before experiment	2·76		2·72		2·47		2·0	
Temperature of secondary before experiment	778° C.		763° C.		670° C.		494 C°.	
Resistance of secondary after experiment	2·75		2·695		2·47		1·94	
Temperature of secondary after experiment	775° C.		754° C.		670° C.		472° C.	
	Magnetising force	Induction per sq. cm.	Magnetising force	Induction per sq. cm.	Magnetising force	Induction per sq. cm.	Magnetising force	Induction per sq. cm.
	0·075	511·8	0·075	{328 / 260	0·075	77	0·075	{54·7 / 35·8
	0·15	1313·9	0·15	{710 / 635	0·15	162	0·15	{98 / 75
	0·3	2482·6	0·3	2304	0·3	427	0·3	{245 / 195
	0·6	3257·4	0·6	5281	0·6	1,516	0·6	{742 / 590
	1·2	3659·2	1·2	6544	2·2	9,381	2·2	9,433
	2·4	4104·0	2·2	7318	7·6	11,562	7·6	12,273
	4·4	4520·0	7·6	8036	47·8	12,859	53·5	15,201
			13·0	8323				
			46·6	8581				

CURVE IX.

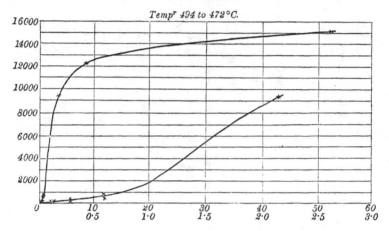

Temp^r 494 to 472°C.

At this stage the ring was allowed to cool down, and on the following day a determination was made of the curve at ordinary temperature of 9°·6 C.   (Curve X.)

Magnetising force

| 0·075 | 0·15 | 0·3 | 0·6 | 1·2 | 2·2 | 4·0 | 6·8 | 11·4 | 17·3 | 57·0 |

Induction per sq. cm.

| 21·6⎫ | 41·1⎫ | 116⎫ | 308⎫ | 1,482 | 6,912 | 10,341 | 12,410 | 13,640 | 14,255 | 15,623 |
| 13·0⎭ | 32·0⎭ | 93⎭ | 273⎭ | | | | | | | |

CURVE X.

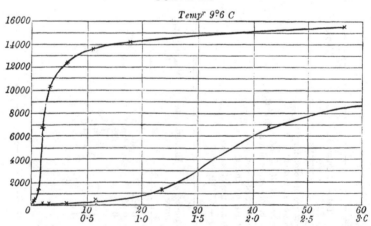

Temp^r 9°6 C

The ring was next heated till the resistance reached about 2·4, was allowed to cool somewhat, and a curve was determined.

(Curve XI.) at a resistance of 1·69 to 1·64. Temperature 378° C.
to 354° C.

Magnetising force

| 0·075 | 0·15 | 0·3 | 0·6 | 1·2 | 2·2 | 4·0 | 7·6 | 13·1 | 51·7 |

Induction per sq. cm.

| 38⎱ 44⎰ | 93⎱ 101⎰ | 263 | 874 | 4,288 | 8,818 | 11,296 | 12,589 | 13,404 | 15,174 |

In addition to the variation of magnetisability depending on
the temperature, these numbers show one or two interesting facts.
Where two observations are given these are the results of successive reversals in opposite directions. After each experiment
the ring was demagnetised by reversals of current; thus currents
successively diminishing in amount were passed through the
primary, each current being reversed ten times. The last
currents gave magnetising forces 1·2, 0·6, 0·3, 0·15, 0·075, 0·05.
The inequality of successive observations is due to the residual
effect of the current last applied; it is remarkable to observe
how greatly this small force affects the result. In Curve XI.
the first deflection was caused by a reversal of a current opposite
to the last demagnetising current.

CURVE XI.

Tempʳ 378 to 354° C.

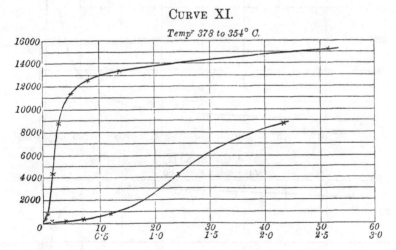

Comparing Curves X. and I. we see that the effect of working
with the sample is to diminish its magnetisability for small forces,
a fact which will be better brought out later.

Referring now to the temperature effects, we see that as the
temperature rises the steepness of the initial part of the curve

13—2

increases, but the maximum magnetisation diminishes. The
coercive force, that is, the force required to completely demagnetise the material after it has been exposed to a great magnetising force, also, judging from the form of the ascending
curves, diminishes greatly.

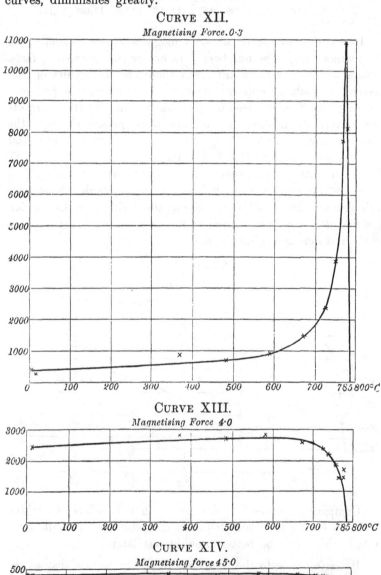

CURVE XII.

Magnetising Force. 0·3

CURVE XIII.

Magnetising Force 4·0

CURVE XIV.

Magnetising force 45·0

In Curves XII., XIII. and XIV. the abscissæ are temperatures, and the ordinates are induction ÷ magnetising force, called by Sir William Thomson the permeability, and usually denoted by μ. These curves correspond to constant magnetising forces of 0·3, 4·0, 45·0. They best illustrate the facts which follow from these experiments. Looking at the curve for 0·3, we see that the permeability at the ordinary temperature is 367 ; that as the temperature rises the permeability rises slowly, but with an accelerated rate of increase ; above 681° C. it increases with very great rapidity, until it attains a maximum of 11,000 at a tempera- ture of 775° C. Above this point it diminishes with extreme rapidity, and is practically unity at a temperature of 786° C.

Regarding the iron as made up of permanently magnetic molecules, the axes of which are more or less directed to parallelism by magnetising force, we may state the facts shown by the curve by saying that rise of temperature diminishes the magnetic moment of the molecules gradually at first, but more and more rapidly as the critical temperature at which the magnetism disappears is approached, but that the facility with which the molecules have their axes directed increases with rise of tempera- ture at first slowly, but very rapidly indeed as the critical temperature is approached.

Whitworth's Mild Steel.—This sample was supplied to me by Sir Joseph Whitworth and Co., who also supplied me with the following analysis of its composition :—

	C	Mn	S	Si	P
Per cent.	·126	·244	·014	·038	·047

The dimensions of the ring were as shown in the accom- panying sketch :

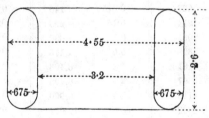

The area of section of the ring is 1·65 sq. cm. The area of the middle line of the secondary coil is estimated to be 2·32 sq. cms.

The secondary coil had 56, the primary 98, convolutions.

The resistance of the secondary and leads was 0·81 at 12° C.

The ring was at once raised to a temperature at which it ceased to be magnetic ; with a magnetising force of 32·0, the total induction was observed to be 80·8, giving the value of the permeability 1·12.

The insulation resistance between the primary and the secondary was observed to be 12,000 ohms.

The ring was now allowed to cool very slowly ; at resistance of 3·00, corresponding to a temperature of 723° C., the ring was non-magnetic; at 2·99, corresponding to 720° C., it was distinctly magnetic.

The following five series of observations were made at descending temperatures, the means of two observations being in each case given ; the sample was demagnetised by reversals after each experiment :—

Table 9, Curve XV.		Table 10, Curve XVI.		Table 13, Curve XIX.	
Resistance at beginning of experiment	2·99	2·71		0·812	
Temperature at beginning of experiment	721°C.	630° C.		12° C.	
Resistance at end of experiment	2·95	2·75		0·812	
Temperature at end of experiment	708° C.	645° C.		12° C.	
Magnetising force	Induction per sq. cm.	Magnetising force	Induction per sq. cm.	Magnetising force	Induction per sq. cm.
0·075	607	0·075	140	0·075	19
0·15	1214	0·15	295	0·15	48
0·3	2031	0·3	1,098	0·3	119
0·6	2698	0·6	4,175	0·6	312
1·2	3181	1·2	6,163	0·9	884
2·2	3607	2·1	8,122	1·7	5,087
7·6	4118	7·5	10,900	3·3	9,535
36·9	4800	38·0	12,074	6·1	12,387
				10·7	13,991
				45·0	16,313

CURVE XV.

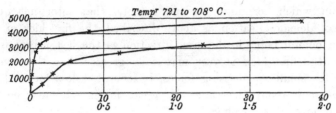

CURVE XVI.

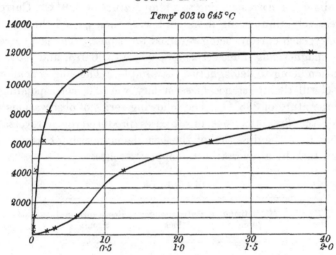

CURVE XIX.

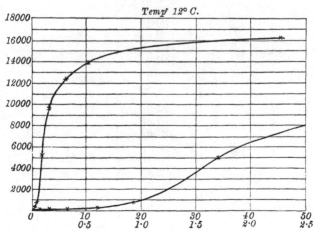

The following experiment is instructive, as showing a pheno-menon which constantly recurs, namely, that after not quite perfect demagnetisation, as above described, the first kick of the galvanometer being in the same direction as the last magnetising force, the first kick is very materially greater than the reverse kick for small magnetising forces, is somewhat less for medium forces, and about the same for great forces. I have no ex-planation of this to offer.

The ring was heated until the resistance of the secondary coil was about 2·4, corresponding to a temperature of 529° C. Currents successively diminishing in amount were then passed through the primary, each current being reversed ten times. The last currents gave magnetising forces 1·2, 0·6, 0·3, 0·15, 0·075, and 0·05, the intention being to demagnetise the sample. The ring was allowed to cool till the resistance of secondary was 2·0, corresponding to a temperature of 398° C. The following series of observations was made: the first kick was in all cases produced by a reversal of current from the direction of the last demagnetising current; the second kick by a reversal in the opposite sense.

TABLE 14.

Magnetising force	Galvanometer kick	Resistance in circuit
0·075	20·5 13·5	12·43
0·15	41·5 32·5	,,
0·3	104·0 81·0	,,
0·6	284·5 241·0	,,
1·2	143·5 150·0	102·43
2·1	262·5 265·0	,,
4·0	351·0 351·0	,,
7·3	210·0 211·5	202·43
12·1	235·5 234·0	,,
43·4	272·5 271·5	,,

The resistance of the secondary coil at the end of the experiment was 2·05; temperature, 415° C.

The sample was again heated until it became non-magnetic, and then allowed to cool very slowly, and the following series of observations were made, the ring being demagnetised as before after each series. The actual kicks of the galvanometer are given, as they illustrate further the point last mentioned. In the first two series only one kick was taken, to save time.

TABLE 15.

Magnetising force	Galvano-meter kick	Resistance in circuit	Induction per sq. cm.	Resistance of coil	Tempera-ture
				3·025	733° C.
0·075	64·5	3·455	61		
0·15	287·0	3·454	273		
0·3	244·0	13·453	903		
0·6	199·0	23·452	1286		
1·2	241·0	23·451	1554		
2·0	290·0	23·450	1870		
				3·019	731

TABLE 16.

Magnetising force	Galvano-meter kick	Resistance in circuit	Induction per sq. cm.	Resistance of coil	Tempera-ture
				3·018	730° C.
0·075	133	13·448	492		
0·15	305	13·448	1128		
0·3	302	23·448	1948		
0·6	91	103·449	2584		
1·2	95	103·449	2698		
37·4	137	103·449	2891		
				3·019	731

TABLE 17.

Magnetising force	Galvano-meter kick	Resistance in circuit	Induction per sq. cm.	Resistance of coil	Tempera-ture
				3·018	730° C.
0·075	214	13·448	792		
0·075	149	13·447	551		
0·075	145	13·445	536		
0·6	102	103·444	2897		
38·4	150	103·442	4260		
				3·012	729

TABLE 18.

Magnetising force	Galvano-meter kick	Resistance in circuit	Induction per sq. cm.	Resistance of coil	Tempera-ture
				3·01	728° C.
0·075	229	13·44	847		
0·075	155	13·44	573		
0·3	89	103·44	2528		
0·075	154	13·43	570		
0·3	96	103·43	2726		
1·2	132	103·43	3749		
7·3	156	103·43	4430		
37·2	181	103·43	5155		
				3·0	725

The sample was again heated until it became non-magnetic. A magnetising force of 0·075 was applied by a current in the primary during heating, and was taken off entirely by breaking the primary circuit when the sample was non-magnetic. The sample was allowed to cool to the ordinary temperature of the room, 12° C., and the following series of observations was made, the first reversal being from the direction of the force of 0·075 which had been applied when the ring was heated.

TABLE 19.

Magnetising force	Galvanometer kick	Resistance in circuit	Induction per sq. cm.
0·075	120	1·244	41
	87	,,	30
0·15	249	,,	85
,,	210	,,	72
0·3	62	11·244	193
	58	,,	179
0·6	178	,,	550
,,	154	,,	476
1·2	59	101·244	1,590
	55	,,	
2·2	227	,,	6,300
	223	,,	
4·0	357	,,	10,080
	363	,,	
7·3	226	201·24	12,553
,,	228	,,	
12·1	252	,,	13,991
	254	,,	
18·8	268	,,	14,876
,,	270	,,	
25·9	275	,,	15,318
	278	,,	
42·4	293	,,	16,148
,,	291	,,	

The ring was now demagnetised by reversed currents, but these were successively reduced to a force of 0·0075, instead of 0·05 as heretofore, and the following series of observations was made:—

TABLE 20.

Magnetising force	Galvanometer kick	Resistance in circuit	Induction per sq. cm.
0·075 ,,	77·0 ⎱ 79·0 ⎰	1·24	27
0·15 ,,	180·0 ⎱ 183·0 ⎰	,,	62
0·3 ,,	52·0 ⎱ 52·5 ⎰	11·24	161
0·6 ,,	126·0 ⎱ 125·0 ⎰	,,	389
1·2 ,,	47·5 ⎱ 47·0 ⎰	101·24	1,314
2·1 ,,	222·0 ⎱ 223·0 ⎰	,,	6,172
4·0 ,,	361·0 ⎱ 366·0 ⎰	,,	10,119
7·5 ,,	228·0 ⎱ 228·0 ⎰	201·24	12,636
12·3 ,,	253·0 ⎱ 252·0 ⎰	,,	13,991
18·8 ,,	270·0 ⎱ 269·0 ⎰	,,	14,903
25·1 ,,	276·5 ⎱ 276·0 ⎰	,,	15,277
42·2 ,,	291·0 ⎱ 289·5 ⎰	,,	16,037

This series shows two things : first, when the demagnetising force is taken low enough there is no asymmetry in the galvanometer kicks; second, the effect of demagnetising by reverse currents is to reduce the amount of induction for low forces.

The ring was now heated to a resistance of secondary of 3·18, temperature 783° C., the ring becoming non-magnetic at 3·03, temperature 734° C. or thereabouts, a magnetising force of about 12 c.g.s. units being constantly applied. The magnetising force was then taken off, and the ring having been allowed to cool, it was magnetised with a force of 46·2.

The ring was again demagnetised, with currents ranging down to 0·0075, and the following series of experiments was made:—

TABLE 22.

Magnetising force	Galvanometer kick	Resistance in circuit	Induction per sq. cm.
0·075 ,,	74·5 ⎱ 76·5 ⎰	1·26	26
0·15 ,,	175·0 ⎱ 180·0 ⎰	,,	62
0·3 ,,	51·5 ⎱ 52·5 ⎰	11·26	161
0·6 ,,	125·0 ⎱ 125·0 ⎰	,,	389
1·2 ,,	231·0 ⎱ 224·0 ⎰	21·26	1,331
2·2 ,,	223·0 ⎱ 224·0 ⎰	101·26	6,272
4·0 ,,	361·0 ⎱ 365·0 ⎰	,,	10,192
7·7 ,,	224·0 ⎱ 229·0 ⎰	201·26	12,576
13·1 ,,	252·0 ⎱ 254·0 ⎰	,,	14,016
20·4 ,,	266·0 ⎱ 269·0 ⎰	,,	14,847
28·8 ,,	277·0 ⎱ 276·0 ⎰	,,	15,346
51·7 ,,	292·0 ⎱ 292·0 ⎰	,,	16,455

It will be seen that this series agrees very closely with Table 20, evidence of the general accuracy of the results.

The ring was lastly demagnetised and heated to a resistance of secondary of 3·19, temperature 787° C., under a magnetising force ·075, which was removed when the ring was at its highest temperature; the ring was cooled, and the following observations made. In this case, however, the first kick was due to a reversal from a current opposed to the current which was applied during heating.

TABLE 23.

Magnetising force	Galvanometer kick	Resistance in circuit	Induction per sq. cm.
0·075 ,,	84·0⎫ 84·5⎭	1·43	33
0·15 ,,	192·0⎫ 195·0⎭	1·43	75
0·3 ,,	60·0⎫ 62·0⎭	11·43	192
0·6 ,,	153·0⎫ 154·0⎭	,,	480
1·2 ,,	321·0⎫ ?302·5⎭	21·43	1,891
2·2 ,,	239·0⎫ 238·0⎭	101·43	6,678
4·0 ,,	367·0⎫ 366·0⎭	,,	10,262
7·3 ,,	227·0⎫ 226·0⎭	201·43	12,576

I have dwelt at length on these experiments because they show that demagnetisation by reversal does not bring back the material to its virgin state, but leaves it in a state in which the induction is much less for small forces and greater for medium forces than a perfectly demagnetised ring would show.

To return to the effects of temperature, Curves XX. and XXI. show the relation of permeability to temperature for magnetising forces 0·3 and 4.

It will be seen that they present the same general characteristics as the curves for wrought-iron. The irregularities are due in part, no doubt, to the dependence of the observations on previous operations on the iron; in part, to uncertainty concerning the exact agreement of temperature of iron and temperature of secondary coil.

Whitworth's Hard Steel.—This sample was supplied to me with the following analysis of its composition :—

	C	Mn	S	Si	P
Per cent.	·962	·212	·017	·164	·016

The dimensions of the ring were exactly the same as the mild steel.

The secondary coil had 56, the primary 101, convolutions.

The resistance of the secondary and leads was ·732 at 8° C.

Experiments were first made with the ring cold, partly to show the changes caused by annealing, and partly to examine the behaviour of the virgin steel.

CURVE XX.

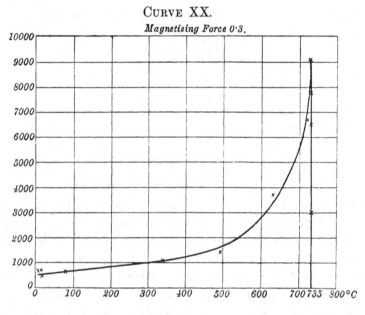

CURVE XXI.

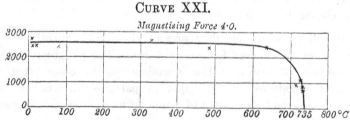

The first series given in Table 24 was made on the virgin steel. The actual elongations on the galvanometer are given, as they afford a better idea of the probable errors of observation. These show that for very small forces the first and second elongations are practically equal, but that for forces between

1 C.G.S. unit and 14 C.G.S. units the first elongation is very materially greater than the later elongations.

The ring was now demagnetised, with magnetising forces ranging down to 0·0045, and the experiment was repeated, the results being shown in Table 25. Comparing them with Table 24, we see that the effect has been to reduce the inductions for low forces, as was the case with mild steel, and to render the kicks practically equal, whether they arise from the current first applied or subsequently applied.

CURVES XXIII., XXIV.

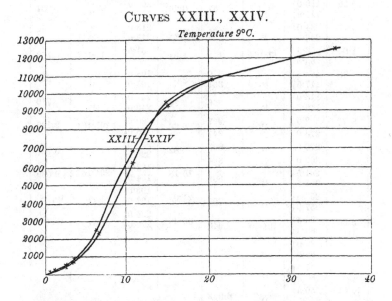

The ring was not now demagnetised; the last current, giving a magnetising force 35·36, was removed, but not reversed, and a series of experiments made, the first reversal in each case being from the direction of the current of 35·36 last applied. The results are given in Table 26.

TABLES 24, 25.

Table 24, Curves XXIII. and XXIV.				Table 25, Curve XXV.			
Magnet-ising force	Galvano-meter kick	Resist-ance in circuit	Induc-tion per sq. cm.	Magnet-ising force	Galvano-meter kick	Resist-ance in circuit	Induc-tion per sq. cm.
0·065	27·0 / 28·0	1·164	9	0·065	26·5 / 26·0	1·164	8
0·13	57·5 / 57·5	,,	18	0·13	55·0 / 53·5	,,	17
0·26	116·0 / 117·5	,,	37	0·26	106·0 / 106·0	,,	34
0·52	234·0 / 236·0	,,	75	0·52	213·0 / 213·0	,,	68
1·04	56·5 / 55·5	11·164	172	1·04	51·5 / 51·5	11·164	158
2·08	123·5 / 117·5	,,	379 / 361	2·08	108·0 / 105·0	,,	328
	116·0 / 116·5	,,	356	3·74	241·0 / 240·0	,,	740
3·74	302·0	,,	927	6·66	80·0 / 78·0	101·164	2,196
	276·0	,,	847	10·82	223·0 / 226·0	,,	6,227
	270·0	,,	829	15·18	163·0 / 164·0	201·164	9,069
	262·0	,,	804	21·0	193·0 / 197·0	,,	10,783
	261·5 / 261·5	,,	802	35·36	226·0 / 226·0	,,	12,498
	258·5 / 257·0	,,	792				
6·66	93·5 / 89·5	101·16	2,543				
	87·0 / 85·0	,,	2,391				
	85·5 / 83·5	,,	2,349				
10·61	250·5 / 247·0	,,	6,922				
	234·5 / 226·0	,,	6,394				
	225·0 / 230·0	,,	6,338				
15·18	168·0 / 171·0	201·16	9,346				
	173·0 / 169·0	,,	9,456				
20·28	190·0 / 197·0 / 194·0 / 193·0	,,	10,728				
35·88	226·0 / 228·0 / 227·0 / 227·0	,,	12,553				

TABLE 26.

Magnetising force	Galvanometer kick	Resistance in circuit	Induction per sq. cm.
0·065	25·0 ⎱ 15·0 ⎰	1·164	8
0·26	111·5	,,	36
	57·0	,,	18
	59·5	,,	19
	57·5	,,	18
3·95	311·5	11·164	956
	140·5	,,	431
	144·0 ⎱ 145·0 ⎰	,,	445
	136·0 ⎱ 132·0 ⎰	,,	411
	135·0 ⎱ 134·0 ⎰	,,	414
11·44	290·0	101·16	8,062
	257·0	,,	7,145
	254·0 ⎱ 251·0 ⎰	,,	7,033
	252·0 ⎱ 250·0 ⎰	,,	6,950
16·43	175·0 ⎱ 173·0 ⎰	201·16	9,622
	172·0 ⎱ 172·0 ⎰	,,	9,512

The ring was now thoroughly demagnetised and heated till it became non-magnetic. It was then cooled slowly, and the following observations were made :—

TABLES 27, 28, 31.

Table 27.		Table 28, Curve XXVIII.	Table 31, Curve XXXI.
Resistance at beginning of experiment	2·805	2·795	2·72
Temperature at beginning of experiment	687° C.	682° C.	657° C.
Resistance at end of experiment	2·795	2·77	2·73
Temperature at end of experiment	682° C.	674° C.	661° C.

Magnetising force	Induction per sq. cm.	Magnetising force	Induction per sq. cm.	Magnetising force	Induction per sq. cm.
0·065	9	0·065	24	0·065	42
0·13	21	0·13	53	0·26	171
0·26	61	0·26	123	1·04	1010
		0·52	291	3·22	3706
		1·04	821	..	..
		2·08	1595	8·32	4885
		3·33	2215	19·8	5708
		5·51	2868		
		8·32	3301		

CURVES XXV., XXVI., XXVII.

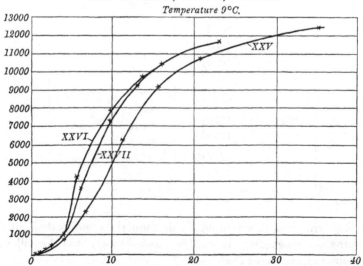

Temperature 9° C.

When cold, the resistance of the secondary coils and leads was
0·768; in calculating the temperatures, it is assumed that the
cold resistance is 0·768.   It is obvious that there is here considerable uncertainty concerning the actual temperatures, owing to the
changes in the condition of the wire due to its oxidation.

The following series was next made, the mean results being
given in

TABLE 34, CURVE XXVI.

Magnetising force	Galvanometer kick	Resistance in circuit	Induction per sq. cm.
0·065	29	1·198	10
0·13	58	1·198	19
0·26	120	1·198	40
0·52	251	1·198	83
1·04	66	11·198	203
3·74	170	21·198	991
6·03	159	101·2	4,420
9·78	283	,,	7,867
13·94	176	201·2	9,733
15·81	187	,,	10,341
22·67	211	,,	11,668

The ring was now demagnetised, and another series of determinations was made, the mean results being given in

TABLE 35, CURVE XXVII.

Magnetising force	Galvanometer kick	Resistance in circuit	Induction per sq. cm.
0·065	26	1·198	9
0·13	54	,,	18
0·26	111	,,	37
0·52	236	,,	78
1·04	60	11·198	185
2·08	132	,,	407
3·74	327	,,	1,007
6·24	130	101·2	3,614
9·78	265	,,	7,367
13·10	168	201·2	9,290
15·7	187	,,	10,341
22·67	211	,,	11,668

Comparing Curves XXV. and XXVII., we see the effect of annealing the iron to be to increase its permeability. Comparing Curves XXVI. and XXVII. we see the effect of demagnetising by reversed currents. Curve XXXIV. shows the relation of permeability to temperature for a force of 1·5.

CURVE XXVIII.

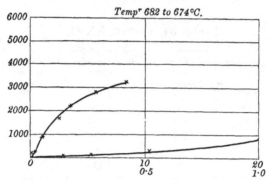

Temp^r 682 to 674°C.

CURVE XXXI.

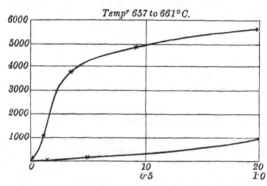

Temp^r 657 to 661° C.

CURVE XXXIV.

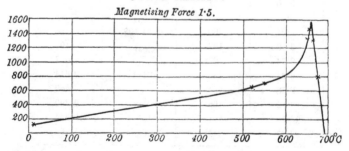

Magnetising Force 1·5.

Manganese Steel.—The sample of this steel was given to me by Mr Hadfield, who also supplied me with the following two analyses of the sample:—

	Per cent.	Per cent.
C	·74	·73
Si	·50	·55
S	·05	·06
P	·08	·09
Mn	11·15	12·06

It is well known that this steel at ordinary temperatures, and for both great and small magnetising forces, is but very slightly magnetic. The object of these experiments was to ascertain whether it became magnetic at any higher temperature.

The dimensions of the ring were as shown in the accompanying section :—

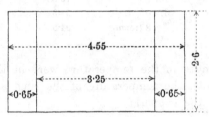

Thus the mean area of section is 1·7 sq. cm., and the mean length of lines of magnetic force 12·3 cms. The ring was wound with 52 convolutions for the secondary and 76 convolutions for the primary. It was not possible to accurately estimate the mean area of the secondary; it is, however, assumed to exceed the mean area of the steel by as much as the secondary of the sample of wrought iron is estimated to exceed the area of that sample; this gives an area of 2·38 sq. cms.

A preliminary experiment at the ordinary temperature gave induction 67·7; magnetising force 26·9.

The induction in the air-space between the wire and steel will be 26·9 × 0·68 = 18·3; deducting this from 67·7, we obtain the induction in the steel equal to 49·4, or 29·0 per sq. cm.; dividing this by 26·9, we obtain 1·08 as the permeability from this experiment.

After the ring had been heated to a high temperature, about 800° C., and had been allowed to cool, a second experiment gave total induction 76, magnetising force 22·8, permeability 1·5.

The ring was again heated and allowed to cool, observations being made both during rise and fall of temperature, with the following results :—

TABLE 36.

Resistance of secondary and leads	Temperature	Total induction	Permeability
	° C.		
0·77	9·0 (room)	67·7	1·08
2·20	476·0	93·1	1·95
3·00	757·0	101·7	2·19
3·23	816·0	71·7	1·45
3·30	841·0	72·0	1·42
3·14	787·0	72·0	1·38
2·80	674·0	92·3	1·99
0·79	8·8 (room)	94·5	1·99

As the changes in the temperature were in this case made somewhat rapidly, the temperature of the ring lags behind the temperature of the copper.

These show: first, that at no temperature does this steel become at all strongly magnetic; second, that at a temperature of a little over 750° C. there is a substantial reduction of permeability ; third, that above this temperature the substance remains slightly magnetic ; fourth, that annealing somewhat increases the permeability of the material.

Resistance of Iron at High Temperatures.

These experiments were made in a perfectly simple way. Coils of very soft iron wire, pianoforte wire, manganese steel wire, and copper wire were insulated with asbestos, were bound together with copper wire so placed as to tend by its conductivity for heat to bring them to the same temperature, and were placed in an iron cylindrical box for heating in a furnace. They were heated with a slowly rising temperature, and the resistance of the wires was successively observed, and the time of each observation noted. By interpolation the resistance of any sample at any time intermediate between the actual observations could be very approximately determined. The points shown in Curves XXXV., XXXVI., XXXVII., were thus determined. In these curves the abscissæ represent the temperatures, and the ordinates the resistance of a wire having unit resistance at 0° C. Curve XXXVII. is manganese steel, which exhibits a fairly constant temperature coefficient of 0·00119; Dr Fleming gives 0·0012 as the temperature coefficient of this material. Curve XXXV. is

CURVE XXXV.

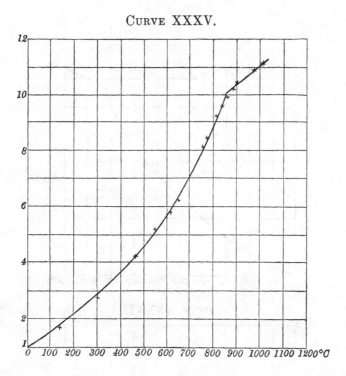

soft iron; at 0° C. the coefficient is 0·0056; the coefficient gradually increases with rise of temperature to 0·019, a little below 855° C.; at 855° C. the coefficient suddenly, or at all events very rapidly, changes to 0·007. Curve XXXVI. is pianoforte wire; at 0° C. the coefficient is 0·0035; the coefficient increases with rise of temperature to 0·016, a little below 812° C.; at 812° C. the coefficient suddenly changes to 0·005. The actual values of the coefficients

above the points of change must be regarded as somewhat uncertain, because the range of temperature is small, and because
the accuracy of the results may be affected by the possible oxidation of the copper. The temperatures of change of coefficient,
855° C. and 812° C., are higher than any critical temperature I
had observed. It was necessary to determine the critical temperatures for magnetisation for the particular samples. A ring was
formed of the respective wires, and was wound with a primary and
secondary coil, and the critical temperature was determined as in

CURVE XXXVI.

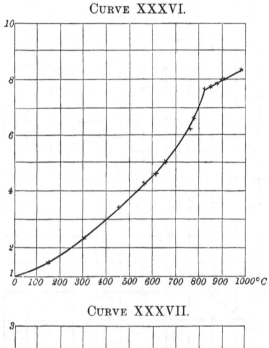

CURVE XXXVII.

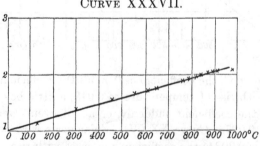

the preceding magnetic experiments: it was found to be for the
soft iron 880° C., for the hard pianoforte wire 838° C. These

temperatures agree with the temperatures of sudden change of resistance coefficient within the limits of errors of observation*.

Some interesting observations were made on the permanent change in the resistance at ordinary temperatures caused in the wires by heating to a high temperature. In the following table are given the actual resistances of wires at the temperature of the room :—

	Before heating	After first heating	Second heating	Third heating
Soft iron	0·629	0·624	0·72	0·735
Pianoforte wire . .	0·851	0·794	0·79	0·74
Manganese steel. .	1·744	1·656	1·61	1·61

In a second experiment the resistances before heating were: soft iron 0·614, pianoforte wire 0·826; after heating, soft wire 0·643, pianoforte wire 0·72.

The effects are opposite in the cases of soft iron and pianoforte wire.

Recalescence of Iron.

Professor Barrett has observed that, if an iron wire be heated to a bright redness and then allowed to cool, this cooling does not go on continuously, but after the wire has sunk to a very dull red it suddenly becomes brighter and then continues to cool down. He surmised that the temperature at which this occurs is the temperature at which the iron ceases to be magnetisable. In repeating Professor Barrett's experiments, I found no difficulty in obtaining the phenomenon with hard steel wire, but I failed to observe it in the case of soft iron wire, or in the case of manganese steel wire. Although other explanations of the phenomenon have been offered, there can never, I think, have been much doubt that

* [*Note added July* 2, 1889.—Sir Joseph Whitworth and Co. have kindly analysed these two wires for me, with the following results:—

	C	Mn	S	Si	P	
Soft iron wire . .	·006	·289	·015	·034	·141	per cent.
Pianoforte wire . .	·724	·157	·010	·132	·030	,, .]

it was due to the liberation of heat owing to some change in the material, and not due to any change in the conductivity or emissive power. This has indeed been satisfactorily proved by Mr Newall*. My method of experiment was exceedingly simple. I took a cylinder of hard steel 6·3 cms. long and 5·1 cms. in diameter, cut a groove in it, and wrapped in the groove a copper wire insulated with asbestos.

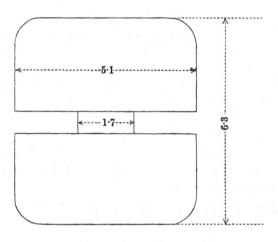

The cylinder was wrapped in a large number of coverings of asbestos paper to retard its cooling; the whole was then heated to a bright redness in a gas furnace; was taken from the furnace and allowed to cool in the open air, the resistance of the copper wire being, from time to time, observed. The result is plotted in Curve XXXVIII., in which the ordinates are the logarithms of the increments of resistance above the resistance at the temperature of the room, and the abscissæ are the times. If the specific heat of the material were constant, and the rate of loss of heat were proportional to the excess of temperature, the curve would be a straight line. It will be observed that below a certain point this is very nearly the case, but that there is a remarkable wave in the curve. The temperature was observed to be falling rapidly, then to be suddenly retarded, next to increase, then again to fall. The temperature reached in the first descent was 680° C. The temperature to which the iron subsequently ascends is 712° C. The temperature at which another sample of hard steel ceased to be

* *Phil. Mag.* June, 1888.

magnetic, determined in the same way by the resistance of a copper coil, was found to be 690° C. This shows that, within the limits of errors of observation, the temperature of recalescence is that at which the material ceases to be magnetic. This curve gives the material for determining the quantity of heat liberated. The dotted lines in the curve show the continuation of the first and second parts of the curve; the horizontal distance between these approximately represents the time during which the material was giving out heat without fall of temperature. After

CURVES XXXVIII., XXXIX.

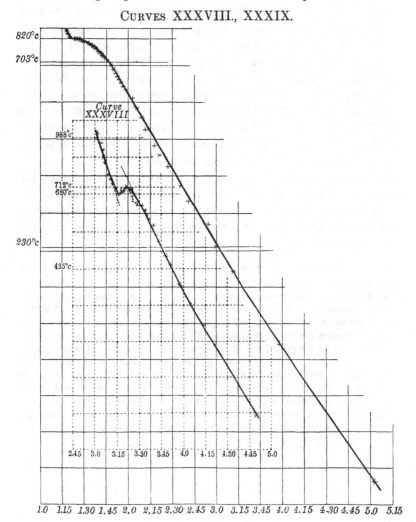

the bend in the curve, the temperature is falling at the rate of
0·21° C. per second. The distance between the two straight parts
of the curve is 810 seconds. It follows that the heat liberated in
recalescence of this sample is 170 times the heat liberated when
the iron falls in temperature 1° C. With the same sample, I have
also observed an ascending curve of temperature. There is, in
this case, no reduction of temperature at the point of recalescence,
but there is a very substantial reduction in the rate at which the
temperature rises*.

A similar experiment was made with a sample of wrought iron
substantially the same as the wrought iron ring first experimented
upon. The result is shown in Curve XXXIX. It will be seen
that there is a great pause in the descent of this curve at a
temperature of 820° C., but that the curve does not sensibly rise.
Determining the heat liberated in the same way as before, we find
the temperature falling after the bend in the curve at the rate of
0·217° C. per second. The distance between the two straight
parts is 960 seconds. Hence, heat liberated in recalescence is 208
times the heat liberated when the iron falls 1° C. in temperature.
The temperature at which a sample ordered at the same time and
place ceased to be magnetic was 780° C. Comparing this result
with that for hard steel, we see that the quantity of heat liberated
is substantially the same, but that in this case there is no material
rise of temperature †.

* [*Note added 2nd July*, 1889.—Some remarks of Mr Tomlinson's suggested
that it might be possible that there would be no recalescence if the iron were
heated but little above the critical point. To test this, I repeated the experiment,
heating the sample to 765°C., very little above the critical point. Curve XXXVIII A.
shows the result. From this it will be seen that the phenomenon is substantially
the same whether the sample is heated to 988°C. or to 765°C.]

† [*Note added 2nd July*, 1889.—In order to complete the proof of the connexion
of recalescence and the disappearance of magnetism, a block of manganese steel
was tried in exactly the same way as the blocks of hard steel and of iron. The result
is shown in Curve XL., from which it will be seen there is no more bend in the
curve than would be accounted for by the presence of a small quantity of magnetic
iron, such a quantity as one would expect from the magnetic results, supposing the
true alloy of manganese and iron to be absolutely non-magnetic.]

Curve XXXVIII a.

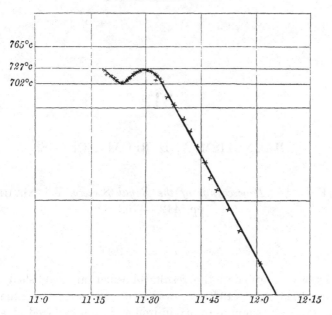

Curve XL.

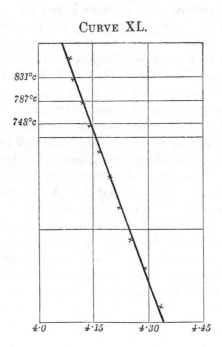

31.

MAGNETISM AND RECALESCENCE.

[From the *Proceedings of the Royal Society*, Vol. XLVIII., pp. 442—446.]

Received October 9, 1890.

IN my experiments the results of which are published, *Phil. Trans.*, 1889, A, p. 443*, I showed that recalescence and the disappearance of magnetisability in iron and steel occurred at about the same temperature. The evidence I then gave was sufficiently satisfactory, but did not amount to absolute proof of the identity of the temperatures. Osmond has shown that the temperature of recalescence depends upon the temperature to which the iron has been heated, also that it differs when the iron is heated and when it is cooled. He also showed that for some sorts of steel the heat is liberated at more than one temperature, notably that in steel with 0·29 per cent. of carbon heat is liberated when cooling at 720° C. and at 660° C., and that with steel with 0·32 per cent. carbon there is a considerable liberation of heat before the temperature is reached when this becomes a maximum. It appeared to be desirable to obtain absolute proof that the change of magnetic property occurred exactly when heat was liberated and absorbed, and to examine, magnetically, Osmond's two temperatures of heat liberation. I have not been able to obtain samples of steel of the size I used, showing two well-marked temperatures of heat liberation and absorption, but I have a ring in which there is liberation of heat extending over a considerable range of temperature.

---

* *Supra*, p. 186.

The samples had the form of rings of the size and shape indicated in Fig. 1. A copper wire was well insulated with asbestos and laid in the groove running round the ring, and was covered with several layers of asbestos paper laid in the groove. This coil was used for measuring temperature by its resistance. The whole ring was served over with asbestos paper and with sheets of mica. The secondary exploring coil was then wound on, next a serving of asbestos paper and mica, and then the primary coil, and, lastly, a good serving of asbestos paper was laid over all. In this way good insulation of the secondary coil was secured, and a reasonable certainty that the temperature coil took the precise temperature of the ring, and that at any time the ring was throughout at one and the same temperature. The whole

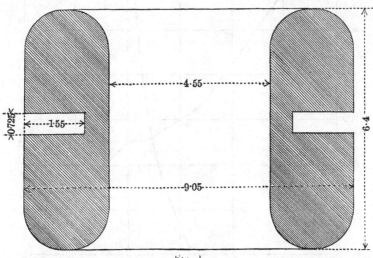

Fig. 1.

was placed in an iron pot, and this again in a Fletcher gas furnace. Observations were made of temperature as the furnace was heating, and from time to time of induction. In each case the time of observation was noted. Similar observations were made as the ring cooled, the furnace being simply extinguished. We are thus enabled to compare directly at the same instant the condition of the same ring as regards magnetism and as regards temperature, and, therefore, qualitatively as regards its absorption or liberation of heat.

In Fig. 2 are the results for a ring containing 0·3 per cent. of carbon or thereabouts. In this case only a cooling curve was taken. It will be observed that there is a considerable liberation of heat, beginning at 2 h. 12 m., temperature 715° C., and continuing to time 2 h. 22 m., temperature 660° C., being apparently somewhat slower at the end. This may, however, be only apparently slower, as the furnace temperature would fall lower in relation to the ring. At time 2 h. 22 m., temperature 660° C., the

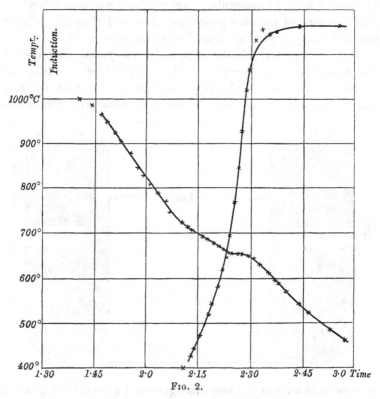

FIG. 2.

rate of liberation becomes much more rapid, so much so that the temperature for a time remains almost stationary. At time 2 h. 29 m. the liberation of heat appears to have ceased and the normal cooling to continue. Now, comparing the kicks of the galvanometer, which are proportional to the induction, we observe that the ring begins to be magnetisable at time 2 h. 12 m., its magnetic property increases till time 2 h. 22 m.; after this point the magnetisability increases much more rapidly, and is practically fully developed at 2 h. 31 m. In this case the development of magnetic property follows precisely the liberation of heat, observed both at the temperature of about 700° C. and at 660° C. We may, therefore, be certain, that both at the higher and lower temperatures of recalescence there is magnetic change, and that the one is as much dependent on the other as the solid condition of ice is upon the liberation of heat when water solidifies. The two changes occur, not only at the same temperature, but simultaneously. A considerable magnetising force, 6·56, was taken, as it was expected and found that the magnetic property would then be more apparent when it was in the intermediate condition between the two temperatures of recalescence.

In Fig. 3 are the results of a ring containing 0·9 per cent. of carbon. In this case we have a curve of heating and of cooling with magnetic properties for comparison, and also a second cooling curve to show the recalescence temperature when the heating had been higher. Unfortunately I had forgot to record the magnetising force: it was, however, much less than in the last case, probably less than unity. Looking at the curve, we see that there is a slight absorption of heat at time 11 h. 17 m., temperature 710° C., with doubtful effect on the magnetism. At time 11 h. 27 m., temperature 770° C., powerful absorption of heat begins and continues to time 11 h. 55 m., temperature 808° C.; it is between these times that the magnetisability is decreasing, and at the latter time that it finally disappears. The heating was continued to about 840° C., and the flame was then put out. In cooling, heat is liberated at one point only, and in this case with a distinct rise of temperature. The recalescence begins at time 3 h. 47 m., temperature 750° C., and it is precisely at this time that the ring begins to be magnetisable. The recalescence continues to time 4 h. 8 m., and at this time, and not before it, the magnetisability practically attains a maximum. Before the last portion of the curve the ring was heated to 966° C. Here no observations were made magnetically. This part of the curve, therefore, only shows the effect of higher heating in lowering the temperature of recalescence.

These experiments show that the liberation and absorption of heat, known as recalescence, and the change in magnetic condition, occur simultaneously. Also that in the case of steel with 0·3 per cent. of carbon both temperatures of liberation of heat are associated with change of magnetic condition.

H. II. 15

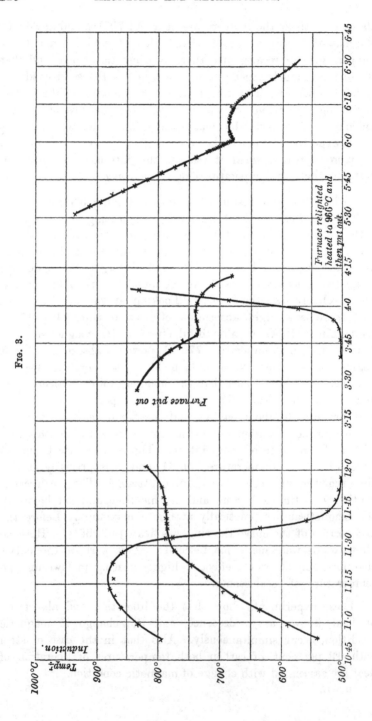

Fig. 3.

32.

MAGNETIC PROPERTIES OF ALLOYS OF NICKEL AND IRON.

[From the *Proceedings of the Royal Society*, Vol. XLVIII. pp. 1—13.]

Received April 17, 1890.

EIGHT different alloys have been examined, distinguished here by the letters of the alphabet. All the samples were given to me by Mr Riley, of the Steel Company of Scotland, who also furnished me with the analysis given with the account of the experiments with each sample.

The methods of experiment were the same as were detailed in my paper on "Magnetic and other Physical Properties of Iron at a High Temperature*." The dimensions of the samples were also the same. For this reason it is unnecessary to recapitulate the methods adopted. I confine myself to a statement of the several results, dealing with each sample in succession.

A. The following is the analysis of this sample :—

Fe.	Ni.	C.	Mn.	S.	P.
97·96	0·97	0·42	0·58	0·03	0·04 per cent.

In this case a magnetisation curve is all that I have obtained free from doubt; the sample was heated and its magnetisation determined at various temperatures for a force of 0·50, but the higher temperatures must be taken as a shade doubtful, as the

---

* *Supra*, p. 186.

secondary broke down before cooling, and I cannot be sure whether or not the resistance of the secondary may have changed.

Table I. gives the results at the ordinary temperature for the material before heating; these are plotted in Curve 1 together with the curve for wrought iron, for comparison.

TABLE I.

Magnetising force.		Induction.
0·06	......	11
0·12	......	29
0·26	......	58
0·53	......	122
1·07	......	303
2·14	......	995
4·7	....	4,560
8·8	......	9,151
16·8	......	12,876
38·9	......	15,651
270·0	......	21,645

CURVE 1.

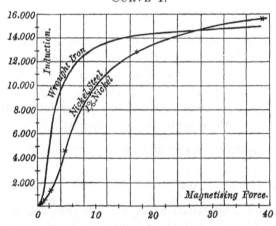

The only noteworthy features are that the coercive force is obviously somewhat considerable, and that the maximum induction is great—greater than that of the more nearly pure iron.

In Curve 2 are shown the results of induction in terms of the temperature for a force of 0·50.

CURVE 2.

l per cent. Nickel. Magnetising Force, 0·50.

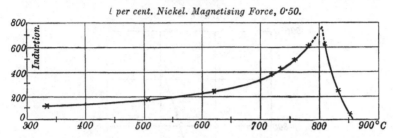

B.   The following is the analysis of the sample :—

Fe.	Ni.	C.	Mn.	S.	P.	Si.
94·799	4·7	0·22	0·23	0·014	0·037	trace per cent.

We have here results of induction in terms of temperature for a magnetising force of 0·12, shown in Curve 3, and for comparison

CURVE 3.

4·7 per cent. Nickel. Magnetising Force, 0·12.

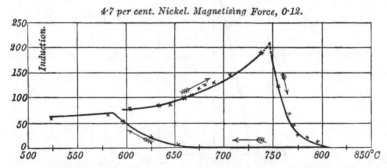

therewith the results of rate of heating and cooling in Curves 4 and 5 respectively.   The experiment with rising temperature was made by simply observing with a watch the hour at which the temperature attained successive values whilst the piece was in the furnace ; the cooling experiments were made in exactly the way described in *Phil. Trans.*, A, 1889, p. 463 ; in the experiment with rising temperature, however (Curve 4), the ordinates are the actual temperatures, not the logarithms of the excess of temperature above the room, as in Curve 5.   The most remarkable feature in Curve 3 is that the material has two critical temperatures, one at which it ceases to be magnetisable with increase of temperature the other, and lower, at which it again becomes magnetisable as the temperatures fall, and that these temperatures differ by about 150° C. Between these temperatures, then, the material can exist

CURVE 4.

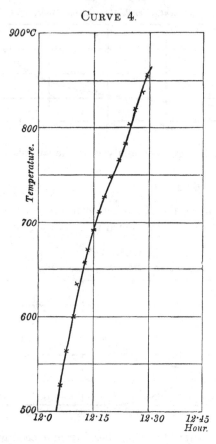

in either of two states—a magnetisable and a non-magnetisable. Note, further, that the curve for decreasing temperature returns into that for increasing temperature, and does not attain to the high value reached when the temperature is increasing. From Curve 4 we see that there is absorption of heat about 750° C., and not before; and from Curve 5 that heat is given off at 632° C., and again at a lower temperature. Comparing these temperatures with Curve 3, it is apparent that the absorption and liberation of heat occur at the same temperature as the loss and return of the capacity for magnetism. From Curve 5 also we may infer that the latent heat liberated in cooling is about 150 times the heat liberated when the temperature of the material falls 1° C. Concerning the latent heat absorbed in heating, nothing can be inferred from Curve 4, excepting the temperature at which it is absorbed.

CURVE 5.

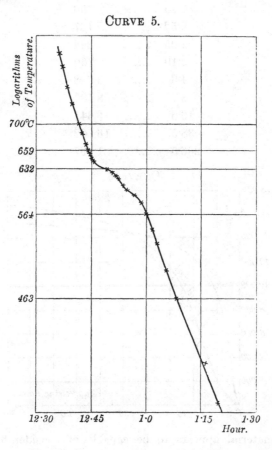

C. This alloy is very similar to the last ; its analysis is—

Fe.	Ni.	C.	Mn.	S.	P.
94·39	4·7	0·27	0·57	0·03	0·04 per cent.

In Table II. are given the results of observations of induction in terms of magnetising force at the ordinary temperature of the room ; and in Curve 6 these are plotted together with the curve for wrought iron.

TABLE II.

Magnetising force.		Induction.
0·06	......	14
0·12	......	29
0·25	......	60
0·52	......	127
1·05	......	294
2·10	......	760
4·6	......	3,068
8·7	......	8,786
16·6	......	13,641
38·5	......	16,702
266·5	......	21,697

CURVE 6.

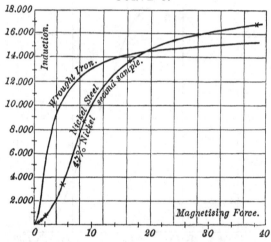

The material appears to be capable of considerably higher magnetisation than wrought iron. In Curve 7 is shown the relation of induction and temperature for two forces, 26·5 and 0·5, the results being obtained on two different days, to the same scale of abscissæ but different scales of ordinates. These curves show the same features as the alloy B, but at a rather lower temperature.

D. This sample contains 22 per cent. of nickel. It was not thoroughly tested, as the supply of CO_2 which happened to be

available was insufficient. Its magnetic properties, however, were
similar to the next sample.

CURVE 7.

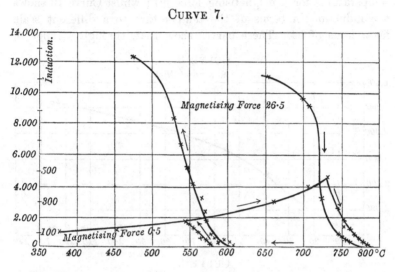

E.   The analysis of this sample was—

Fe.	Ni.	C.	Mn.	S.	P.	Si.
74·31	24·5	0·27	0·85	0·01	0·04	0·02 per cent.

As the material was given to me it was non-magnetisable at
ordinary temperature ; that is to say, the permeability was small,
about 1·4, and the induction was precisely proportional to the
magnetising force. The ring on being heated remained non-
magnetisable up to 700° C. or 800° C. A block of the material
did not recalesce on being heated to a high temperature and being
allowed to cool.

On being placed in a freezing mixture, the material became
magnetic at a temperature a little below freezing point.

The material was next cooled to a temperature of about
− 51° C. by means of solid carbonic acid. After the temperature
had returned to 13° C. the curve of magnetisation was ascertained
as shown in Curve 8; from this it will be seen that the ring of
the material which was previously non-magnetisable at 13° C. is
now decidedly magnetisable at the same temperature. On heating
the material, it remained magnetisable until it reached a temperature of 580° C. At this temperature it became non-magnet-

isable, and, on cooling, remained non-magnetisable at the ordinary temperature of the room. Curve 9 shows the induction at various temperatures for a magnetising force 6·7; whilst Curve 10 shows the induction in terms of the temperature to a different scale for a force of 64. These curves show that, through a range of

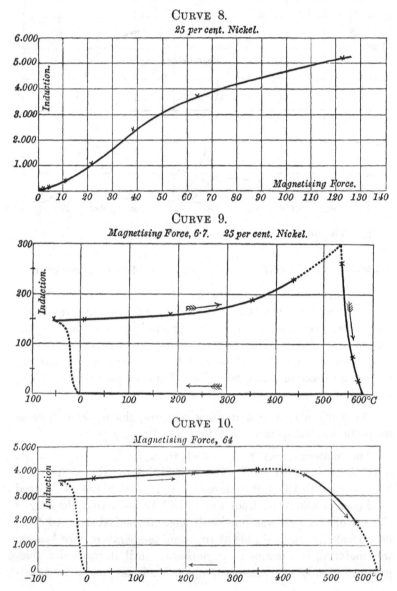

CURVE 8.

25 per cent. Nickel.

CURVE 9.

Magnetising Force, 6·7. 25 per cent. Nickel.

CURVE 10.

Magnetising Force, 64

temperature from somewhat below freezing to 580° C., this material exists in two states, either being quite stable, the one being non-magnetisable, the other magnetisable. It changes from non-magnetisable to magnetisable if the temperature be reduced a little below freezing; the magnetisable state of the material does not change from magnetisable to non-magnetisable until the temperature is raised to 580° C.

The same kind of thing can be seen in a much less degree with ordinary steel. Over a small range this can exist in two states; but in changing its state from non-magnetisable to magnetisable a considerable amount of heat is liberated, which causes rise of temperature in the steel. It is observed in samples B and C of nickel steel, as we have just seen, but at a higher temperature.

As might be expected, the other physical properties of this material change with its magnetic properties. Mr Riley has kindly supplied me with wire.

The wire as sent to me was magnetisable as tested by means of a magnet in the ordinary way. On heating it to a dull redness it became non-magnetisable, whether it was cooled slowly or exceedingly rapidly, by plunging it into cold water. A quantity of the wire was brought into the non-magnetisable state by heating it and allowing it to cool. The electric resistance of a portion of this wire, about 5 metres in length, was ascertained in terms of the temperature; it was first of all tried at the ordinary temperature, and then at temperatures up to 340° C. The specific resistances at these temperatures are indicated in Curve 11 by the numbers 1, 2, 3. The wire was then cooled by means of solid carbonic acid. The supposed course of change of resistance is indicated by the dotted line on the curve; the actual observations of resistance, however, are indicated by the crosses in the neighbourhood of the letter A on the curve. The wire was then allowed to return to the temperature of the room, and was subsequently heated, the actual observations being shown by crosses on the lower branches of the curve, the heating was continued to a temperature of 680° C., and the metal was then allowed to cool, the actual observations being still shown by crosses. From this curve it will be seen that in the two states of the metal (magnetisable and non-magnetisable) the resistances at ordinary temperatures are quite different. The specific resistance in the magnetisable condition is about 0·000052; in the non-magnetisable condition it is about 0·000072. The curve of resistance in terms of the temperature of the material in the magnetisable condition has a close resemblance to that of soft iron, excepting that the coefficient of variation is much smaller, as, indeed, one would expect in the case of an alloy; at 20° C. the coefficient is about 0·00132; just below 600° C. it is about 0·0040,

CURVE 11.

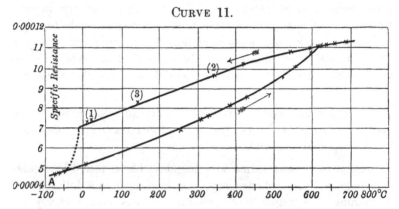

and above 600° C. it has fallen to a value less than that which it had at 20° C. The change in electrical resistance effected by cooling is almost as remarkable as the change in the magnetic properties.

Samples of the wire were next tested in Professor Kennedy's laboratory for mechanical strength. Five samples of the wire were taken which had been heated and were in the non-magnetisable state, and five which had been cooled and were in the magnetisable state. There was a marked difference in the hardness of these two samples; the non-magnetisable was extremely soft, and the magnetisable tolerably hard. Of the five non-magnetisable samples the highest breaking stress was 50·52 tons per square inch, the lowest 48·75; the greatest extension was 33 per cent., the lowest 30 per cent. Of the magnetisable samples, the highest breaking stress was 88·12 tons per square inch, the lowest 85·76; the highest extension was 8·33, the lowest 6·70. The broken fragments, both of the wire which had originally been magnetisable and that which had been non-magnetisable, were now found to be magnetisable. If this material could be produced at a lower cost,

these facts would have a very important bearing. As a mild
steel, the non-magnetisable material is very fine, having so high
a breaking stress for so great an elongation at rupture. Suppose
it were used for any purpose for which a mild steel is suitable on
account of this considerable elongation at rupture, if exposed to a
sharp frost its properties would be completely changed—it would
become essentially a hard steel, and it would remain a hard steel
until it had actually been heated to a temperature of 600° C.

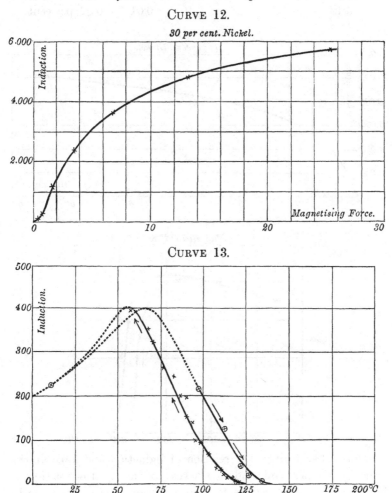

CURVE 12.

30 per cent. Nickel.

CURVE 13.

F. This sample contains 30 per cent. of nickel. Curve 12
shows the relation of induction to magnetising force at the ordinary

temperature, and Curve 13 the relation of induction and temperature for a force of 0·65. The remarkable feature here is the low temperature at which the change between magnetisable and non-magnetisable occurs, whether the temperature is rising or falling. Comparing it with the last sample, we see that the character of the material with regard to magnetism is entirely changed.

G.   The analysis of this sample is—

Fe.	Ni.	C.	Mn.	S.	P.
66·19	33·0	0·28	0·50	0·01	0·02 per cent.

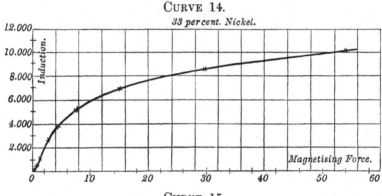

CURVE 14.

33 per cent. Nickel.

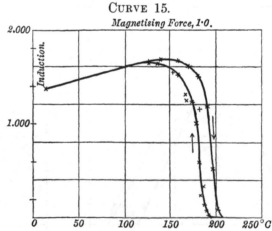

CURVE 15.

Magnetising Force, 1·0.

In Curve 14 is given the relation of induction and force at the ordinary temperature, and in Curves 15 and 16 the relation of induction and temperature for forces 1·0 and 30·3. The remarkable feature of this material is the complete difference from the last but one, and the low temperature of change. There is but very little difference between the temperatures of change when heated and when cooled.

CURVE 16.

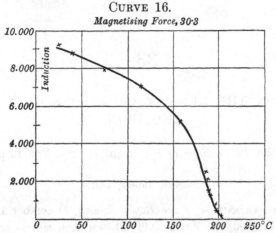

Magnetising Force, 30·3

H.   The analysis of this sample, as furnished by Mr Riley is—

Fe.	Ni.	C.	Mn.	S.	P.
26·50	73·0	0·18	0·30	0·01	0·01 per cent.

In Curve 17 is given the relation of induction and force at the ordinary temperature.   It is curious to remark that the induction

CURVE 17.

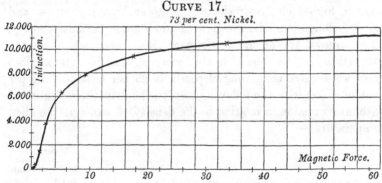

73 per cent. Nickel.

for considerable forces is greater than in the steel with 33 per cent. of nickel, and that it is greater than for a mechanical mixture of iron and nickel in the proportions of the analysis, however the particles might be arranged in relation to each other.

The critical temperature of the material is 600° C.; it shows no material difference between the critical temperatures for increasing and diminishing temperatures.

33.

NOTE ON THE DENSITY OF ALLOYS OF NICKEL AND IRON.

[From the *Proceedings of the Royal Society*, Vol. L. p. 62.]

Received June 3, 1891.

IN the *Proceedings of the Royal Society*, December 12, 1889, January 16, 1890, and May 1, 1890*, I described certain properties of alloys of nickel and iron containing respectively 22 per cent. and 25 per cent. of nickel. These alloys can exist in two states at temperatures between 20° or 30° C. below freezing and a temperature of near 600° C. After cooling, the alloys are magnetisable, have a lower electric resistance, a higher breaking stress, and less elongation; after heating, the alloys are not magnetisable, have a higher electric resistance, a lower breaking stress, and greater elongation. I have now to add another curious property. These alloys are about 2 per cent. less dense when in the magnetisable than when in the non-magnetisable state. Two rings were tested containing respectively 25 per cent. and 22 per cent. of nickel with the following results, the densities being given without correction in relation to the density of water at the then temperature:—

	Nickel, 25 per cent.		Nickel, 22 per cent.	
	Density	Temp.	Density	Temp.
After heating, non-magnetisable	8·15	15·1	8·13	16·5
After cooling, magnetisable	7·99	14·5	7·96	15·6
After heating again, non-magnetisable	8·15	18·0	8·12	18·2
After cooling again, magnetisable	7·97	22·0	7·95	21·8

The rings were each time cooled to from − 100° C. to − 110° C. by carbonic acid and ether *in vacuo*.

* *Supra*, p. 227.

34.

MAGNETIC PROPERTIES OF PURE IRON. By FRANCIS
LYDALL and ALFRED W. S. POCKLINGTON. Communicated
by J. HOPKINSON, F.R.S.

[From the *Proceedings of the Royal Society*, Vol. LII.
pp. 228—233.]

Received May 4,—Read June 16, 1892.

THE following results were obtained at King's College, Strand,
for a specimen of very pure iron. The experiments were made
under the direction of Dr Hopkinson. The sample was supplied
to him by Sir Frederick Abel, K.C.B., F.R.S., to whom it was sent
by Colonel Dyer, of the Elswick Works. It is of almost pure
iron, and the substances other than iron are stated to be :—

Carbon.	Silicon.	Phosphorus.	Sulphur.	Manganese.
Trace.	Trace.	None.	0·013	0·1

The method of experiment is the same as that described in
Dr Hopkinson's paper before this Society on the "Magnetisation
of Iron at High Temperatures," viz., taking a curve of induction
at the temperature of the atmosphere, and then at increasing
temperatures until the critical point is reached. The temperatures, as in his paper, are calculated from the resistances of
the secondary winding, the increase of resistance per 1° C.
being assumed to be 0·00388 of the resistance at 20° C. In

H. II. 16

brackets are also given the temperatures calculated by Benoit's formula—

Resistance at $t°$ C. = resistance at $0°$ C. $\{1 + 0.00367t$
$$+ 0.000000587t^2\}\,*.$$

The dimensions of the iron ring are—

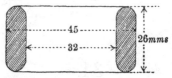

as in the earlier experiments.

Fig. 1 gives the curve of induction taken at $10.5°$ C. compared with the sample of wrought iron of Dr Hopkinson's paper, just referred to, taken at $8.5°$ C. It shows the very high induction developed in the pure specimen for a moderate magnetising force, and also the small amount of hysteresis. The following are the actual values of induction, B, and magnetising force, H :—

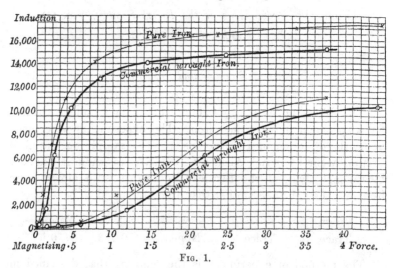

FIG. 1.

Resistance of secondary $= 0.75$ ohm.   Temperature, $10.5°$ C. (pure specimen, marked ×).

B	34	118	467	2700	7060	10,980	14,160	15,590	16,570	17,120	17,440
H	0.15	0.38	0.6	1.06	2.11	3.77	7.48	13.36	23.25	33.65	44.66

* Everett's *C.G.S. Units and Physical Constants*, p. 160.

Temperature, 8·5° C. (ordinary specimen, marked ◯).

B ...	39·5	116	329	1560	6041	10,144	12,633	14,059	14,702	15,149
H ...	0·15	0·3	0·6	1·2	2·2	4·4	8·2	14·7	24·7	37·2

Figs. 2, 3, 4, and 5 respectively give the curves taken at the following temperatures, as calculated from the secondary resistances—658° (676°), 727° (738°), 770° (780°), 855° (857°).

The values for these curves are as follows :—

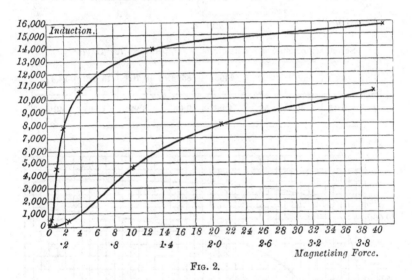

Fig. 2.

Secondary resistance = 2·706 ohms.   Temperature, 658° C. (676°).

B	103·37	360	4453	7899	10,556	13,836	15,640
H	0·09	0·25	1·02	2·08	3·97	12·96	40·92

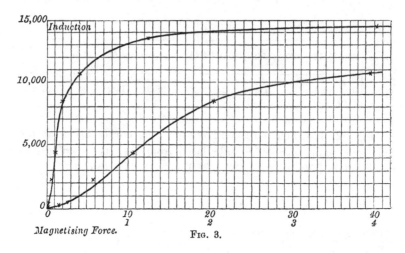

FIG. 3.

Secondary resistance = 2·91 ohms.   Temperature, 727° C. (738°).

B ...	167	532	2260	4405	8553	10,763	13,580	14,442
H ...	0·15	0·28	0·56	1·08	2·05	3·97	12·62	40·4

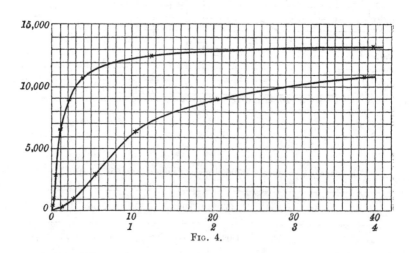

FIG. 4.

Secondary resistance = 3·046 ohms.   Temperature, 770° C. (780°).

B ...	249	1030	2971	6441	8944	10,727	12,528	13,139
H ...	0·14	0·28	0·56	1·07	2·08	3·87	12·6	39·7

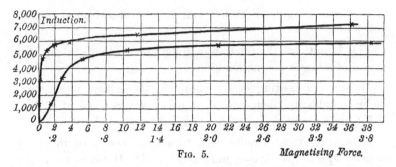

FIG. 5.

Secondary resistance = 3·303 ohms.   Temperature, 855° C. (857°).

B ...	1316	3123	4682	5347	5779	5902	6513	7139
H ...	0·15	0·28	0·53	1·05	2·08	3·87	11·9	36·1

In these we see for a rise of temperature a marked decrease of hysteresis and a very much lower maximum of induction.

Also that for a small magnetising force the permeability rises very remarkably with the temperature, but just the reverse for a force of, say, "40."

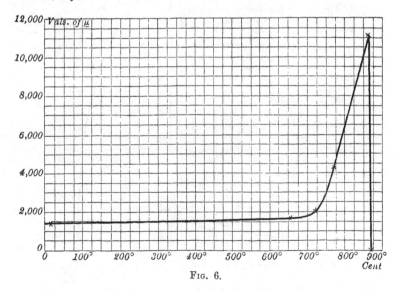

FIG. 6.

Fig. 6 shows the rise of permeability in relation to temperature when $H = 0·3$, the maximum permeability observed being 11,100 for a magnetising force of 0·3, and at a temperature of 855° C. (857°).

Fig. 7 contrasts the relation of induction to temperature at a small and a larger magnetising force.

During the heating of the specimen, the critical point, when the iron suddenly became non-magnetic, was reached at 874° C. (875°), and on cooling it became magnetic at 835° C. (838°).

Comparing these results with those obtained with the more ordinary specimens of iron mentioned in Dr Hopkinson's paper, we have here 874° C. as against 786° C., while in an experiment on some soft iron wire the critical temperature was 880° C., and for hard pianoforte wire it was 838° C.

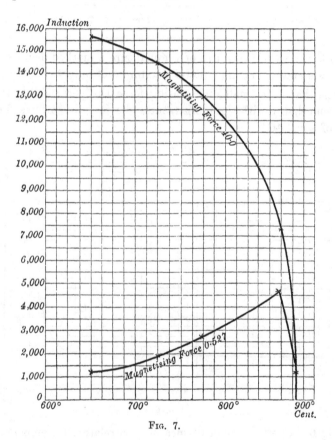

FIG. 7.

35.

MAGNETIC VISCOSITY. By J. HOPKINSON, F.R.S., and
B. HOPKINSON.

[From *The Electrician*, September 9, 1892.]

THE experiments herein described were made in the Siemens
Laboratory at King's College, London. The object was to ascertain
whether the cyclical change in the magnetic induction in iron
due to a given cyclical change in the magnetising force is
independent of the speed at which the change is effected, that is,
whether any sign of "magnetic viscosity" or "magnetic lag" can
be observed when the rate of change is such as is found in transformers. The question is one of much practical interest, and has
been much discussed, amongst others, by Prof. Ewing at the
recent meeting of the British Association. Prof. Ewing has
devised apparatus adapted to deal with this matter as well as
for drawing curves of magnetisation.

We have experimented on two materials. One was soft iron
and the other a hard steel containing about 0·6 per cent. of carbon.
Both samples were supplied by Messrs Richard Johnson.

It had been found, in experiments on ordinary transformers,
that the local currents in the iron made it impossible to form
a correct estimate of the magnetising force. The effect of such
local currents can, of course, be diminished by using finer wire or
plates and better insulation. Our material was in the form of
wire $\frac{1}{100}$ in. diameter, and the wire was varnished with shellac to
insure insulation. It was wound into a ring having a sectional
area of 1·04 sq. cm. in the case of soft iron, and 1·08 sq. cm. in the

case of hard steel, and about 9 cms. in diameter.   The ring was
wound with about 200 turns of copper wire, and with a fine wire
for use with the ballistic galvanometer.   An inspection of the
curves showing the results will satisfy the reader that the effects
of local currents were negligible.

For determining the points on the closed curve of magnetisation, given by rapid reversals of the current in the coil, the ring
was connected in series with a non-inductive resistance to the
poles of an alternate-current generator or a transformer excited
by the generator, thus :—

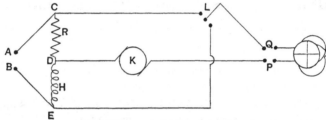

Diagram of connexions.

in which AB are the poles of the transformer or generator, CD
the terminals of the non-inductive resistance R, H the coil
surrounding the ring, P and Q the studs of a reversing key connected to the quadrant of a Thomson quadrant electrometer, L a
key by means of which Q could be connected with C or E at will,
and K a revolving contact maker through which P was connected
to D.   A condenser was connected to P and Q in order to steady
the electrometer readings.   The contact maker K was bolted on
to the axle of the generator.   It consists of a circular disc of
ebonite, about 13 in. in diameter, having a small slip of copper
about $\frac{1}{16}$ in. wide let into its circumference.   A small steel brush
presses on the circumference and makes contact with the piece of
copper once in every revolution.   The position of the brush can
be read off on a graduated circle, and thus contact can be made at
any desired instant in the revolution, and that instant determined
by means of the graduated circle.   The quadrant electrometer
thus gives the instantaneous value of the difference of potential
between the points C and D, or the points D and E, according to
the direction of the key L.   The frequency was in all cases,
except one, 125 complete periods per second.   From observations
of the values of the potential difference between C and D at

different times in the period, a curve (*A*) was plotted giving the current or magnetising force in terms of the time; a similar curve (*B*) was plotted for the electromotive force between *D* and *E*. The curve (*B*) corrected by subtracting the electromotive force due to the resistance of the coil *H* gives the potential or time rate of variation of the induction in terms of the time. Hence, the area of *B* up to any point plus a constant, is proportional to the induction corresponding to that point. This is shown in curve *C*, which is the integral of *B*. A fourth curve *D* was then plotted in which the abscissa of any point is proportional to the magnetising force at any time (got from curve *A*) and the ordinate is proportional to the induction at the same time (got from curve *C*).

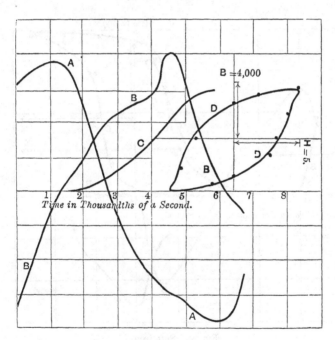

Fɪɢ. 1. Soft iron. Frequency 103.

It is obvious that at the point where *B* cuts the axis the induction is a maximum; hence, if there were no "magnetic lag" and no currents in the iron, this point should occur at the same time as that at which the current is a maximum. In the curves referred to this is seen to be nearly the case.

The slow cycles were obtained from the ballistic galvanometer, by observing the throw due to a known sudden change in the magnetic force.   Care was taken always to take the material through the same cycle.   The points got by the slow method are in each case shown in absolute measure on the same scale as that to which the quick curves are drawn, and are indicated by black dots; they are hardly numerous enough to draw a curve with certainty, but are ample to exhibit the identity of or the character of the difference, if any, between the curves, as determined by the two methods.

Figs. 1 and 2 show the results of experiments on soft iron, Figs. 3, 4, and 5 were obtained from the hard steel.   In all these the agreement between the slow and rapid cycles is fairly close.

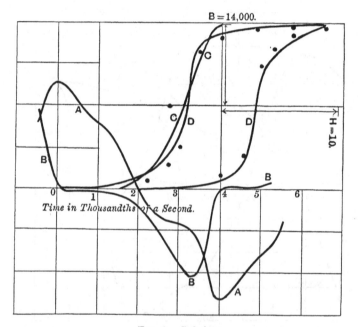

Fig. 2.   Soft iron.

Fig. 6 is interesting, as showing the large effect of local currents. It was obtained from the same sample of steel wire as Figs. 3, 4, and 5, but the wire was *not* varnished.   It will be seen that the maximum induction lags behind the maximum magnetising force about one-sixth of a complete period, and also that the maximum

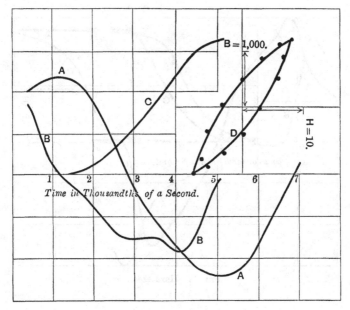

FIG. 3. Hard steel.

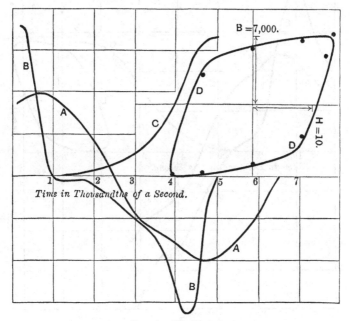

FIG. 4. Hard steel.

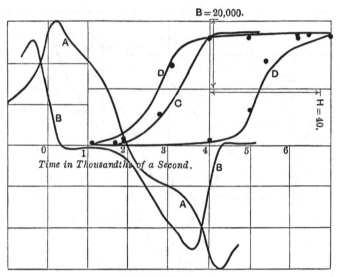

FIG. 5. Hard steel.

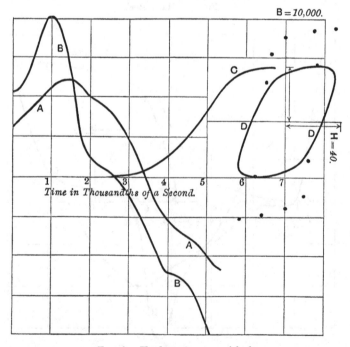

FIG. 6. Hard steel, unvarnished.

induction attained is but 10,000 as against 17,200 obtained from the same (apparent) magnetising force by the slow method.

Hence the general result is, that up to the frequency tried, *i.e.*, about 125 per second, there is no sign of magnetic viscosity; the magnetic cycle is unaffected by the frequency so far as the maximum induction for a given magnetising force is concerned; but that there is a sensible difference between the curve as determined by the two methods, most apparent in that part of the curve preceding the maximum induction. This difference is well shown in Fig. 5. We have not yet fully investigated this feature; possibly it arises from something peculiar to experiments with the ballistic galvanometer.

36.

MAGNETIC VISCOSITY. By J. Hopkinson, D.Sc., F.R.S., E. Wilson, and F. Lydall.

[From the *Proceedings of the Royal Society*, Vol. LIII. pp. 352—368.]

Received March 8, 1893.

THE following experiments were carried out in the Siemens Laboratory, King's College, London, and are a continuation of experiments by J. Hopkinson and B. Hopkinson, a description of which appeared in the *Electrician*, September 9, 1892 *.

In that paper determinations were given of curves showing the relation between the induction and the magnetising force, for rings of fine wire of soft iron and steel, through complete cycles with varying amplitudes of magnetising force, both with the ordinary ballistic method and with alternating currents of a frequency up to 125 complete periods per second. It was shown that if the induction was moderate in amount, for example, 3000 or 4000, the two curves closely agreed; but, if the induction was considerable, for example, 16,000, the curves differed somewhat, particularly in that part of the curve preceding the maximum induction. The difference was greater with steel than with soft iron.

It was not then determined whether this difference was a true time effect or was in some way due to the ballistic galvanometer. The present paper is addressed to settling this point.

* *Supra*, p. 247.

The ring to which the following experiments refer is of hard steel containing about 0·6 per cent. of carbon, in the form of wire $\frac{1}{100}$ in. diameter, varnished with shellac to insure insulation. The material was supplied by Messrs Richard Johnson. The ring is about 9 cm. diameter, and has a sectional area of 1·08 sq. cm.; it is wound with 200 turns of copper wire, and with 80 turns of fine wire for use with the ballistic galvanometer.

In the *Electrician* paper the static curve of hysteresis was determined by the ballistic galvanometer, the connexions being made according to the diagram in Fig. 1: where R is the hard steel wire ring, B is the ballistic galvanometer, S_1 is a reversing switch, and S_2 is a small short-circuiting switch for the purpose of suddenly inserting a resistance R_1 into the primary circuit. The resistance R_2 was so adjusted that the maximum current in the primary circuit was such as to give the desired maximum magnetising force on the ring.

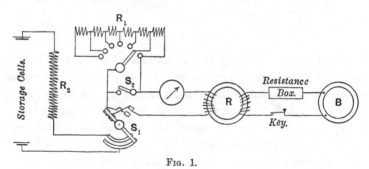

Fig. 1.

In taking the kicks on the ballistic galvanometer the method adopted was as follows:—Having closed the primary by means of S_1, the switch S_2 was suddenly opened, thus allowing the magnetising force to drop to an amount determined by R_1, and the kick observed. A total reversal was then taken with S_1, and the kick again observed. The closing of S_2 again brought up the magnetising force to its maximum in the opposite direction to that at starting.

In a letter to the editor of the *Electrician*, September 16, 1892, Mr Evershed stated that, "Had the slow cycle been obtained by the method described by Mr Vignoles*, Messrs Hopkinson

* *Electrician*, May 15 and 22, 1891.

would have found it in almost absolute agreement with the quick cycle curve."

To settle this point the static curve of hysteresis was obtained by the ballistic galvanometer, the connexions being made according to the diagram in Fig. 2. This is not the method of experiment alluded to by Mr Evershed, but it is capable of varying the magnetising force in the same way as is described by him. R is the hard steel wire ring, B is the ballistic galvanometer, S_1 is a reversing switch, and S_2 a small switch for the

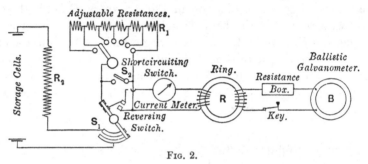

Fig. 2.

purpose of short-circuiting the adjustable resistance R_1. The difference between this diagram and that in Fig. 1 is that R_1 can be suddenly inserted into the primary circuit by one stroke of the reversing switch S_1. In this way it is possible to vary the magnetising force from one maximum through zero to any desired point within the other maximum by one motion of the switch S_1: which operation takes but a small fraction of a second to perform.

In Fig. 3 the points marked × were obtained by the method in Fig. 1; the points marked • being obtained by the method in Fig. 2. Table I. gives the values for B and H, from which these points have been plotted, and their close agreement proves that the difference found between the static and quick cycle curves is not due to the cause suggested by Mr Evershed. In each case the battery used had a potential difference of 108 volts, the periodic time of the ballistic needle being 10 seconds.

It was observed, when taking the hysteresis curve by the method in Fig. 2, that the sum of the inductions found by varying the magnetising force from one maximum to an intermediate point, and from that point to the other maximum, did not exactly equal the induction got by varying the magnetising force direct from one maximum to the other.

To investigate this with the ballistic galvanometer the magnetising force (Fig. 3) was taken from one maximum through zero to the point *a* by one motion of the reversing switch handle, and the galvanometer circuit closed at known intervals of time *after* such change, the deflection being noted. This deflection does not represent an impulsive electromotive force, nor yet a

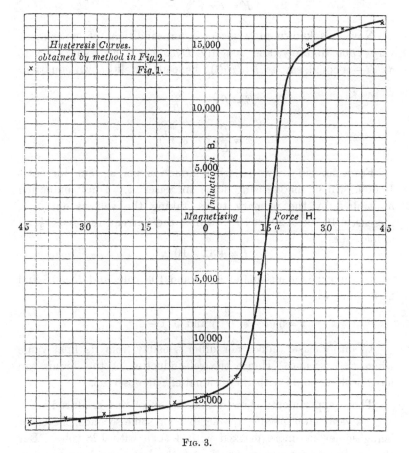

FIG. 3.

constant current, but is caused by a current through the galvanometer diminishing in amount somewhat rapidly. It might arise from the comparatively slow rate at which the magnetising current changes, owing to the self-induction of the circuit, or it might arise from a finite time required to develop the induction corresponding to a given magnetising force. The former would be readily calculable if the ring had a definite self-induction; in our case it is approximately calculable.

Let R be resistance of primary circuit, E the applied electromotive force, x the current, and I the total induction multiplied by the number of primary turns:

$$E = Rx + \frac{dI}{dt}.$$

Now I is known in terms of x for conditions of experiment very approximately, and roughly dI/dt has a constant ratio to dx/dt—is equal, say, to $L\,(dx/dt)$; hence the well-known equation

$$E = Rx + L\frac{dx}{dt},$$

$$x = \frac{E}{R}(1 - \epsilon^{-\frac{R}{L}t}).$$

From our curves we see that induction per sq. cm. increases 10,000, whilst magnetising force increases 4. Total induction multiplied by the primary turns, taking the volt as our unit, increases $10,800 \times 200 \times 10^{-8}$, whilst the current increases $\frac{1}{2}$ an ampère, i.e.,

$$L = 4\cdot32 \times 10^{-2}.$$

In the experiments made $E = 4$ and 108 volts and $R = 0\cdot8$ and $21\cdot6$ ohms, whence

$$x = 5\,(1 - \epsilon^{-\frac{80}{4\cdot32}t}) \text{ and } 5\,(1 - \epsilon^{-\frac{2160}{4\cdot32}t}).$$

In either case x does not differ sensibly from its final value when $t = \frac{2}{5}$ second. Hence the self-induction of the circuit can have nothing to do with the residual effects observed.

These experiments showed that an effect was produced upon the galvanometer needle, appreciable for some seconds, the effect being somewhat more marked with 4 than with 108 volts. But the whole amount was so small as to be less than 1 per cent. of the total change of induction; from which we infer that no material difference exists between curves of induction determined by the ballistic galvanometer and the inductions caused by magnetising forces operating for many seconds.

Effect of tapping the Specimen.—Having taken the magnetising force from its maximum through zero to the point *a* as before, the effect of tapping was marked, especially in the case of soft iron, when a kick corresponding to an acquirement of 633 lines of induction per sq. cm. was observed.

The following experiments on the hard steel-wire ring were carried out with the alternator, the object being to ascertain if a time effect on magnetism exists. The ballistic curve (Fig. 3) has been taken as a standard with which to compare the respective hysteresis curves. In each case the maximum magnetising force has been made as nearly as possible to agree with that used when taking the ballistic curve, and the method of test was that employed in the *Electrician* paper. For the sake of completeness the diagram, Fig. 4, and description are given over again.

Quoting from that paper, we have: "For determining the points on the closed curve of magnetisation, given by rapid reversals of the current in the coil, the ring was connected in series with a non-inductive resistance to the poles of an alternate-current generator, or a transformer excited by the generator, thus :—

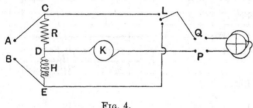

FIG. 4.

in which *A*, *B* are the poles of the transformer or generator ; *C*, *D* the terminals of the non-inductive resistance *R* ; *H* the coil surrounding the ring; *P* and *Q* the studs of a reversing key connected to the quadrant of a Thomson quadrant electrometer ; *L* a key by means of which *Q* could be connected with *C* or *E* at will; and *K* a revolving contact maker, through which *P* was connected to *D*. A condenser was connected to *P* and *Q* in order to steady the electrometer readings. The contact maker *K* was bolted on to the axle of the generator. It consists of a circular disc of ebonite, about 13 in. in diameter, having a small slip of copper, about $\frac{1}{16}$ in. wide, let into its circumference. A small steel brush presses on the circumference, and makes contact with the piece of copper once in every revolution. The position of the

17—2

brush can be read off on a graduated circle. The quadrant electrometer thus gives the instantaneous value of the difference of
potential between the points C and D, or the points D and E,
according to the direction of the key L."

Frequencies of 5, 72, and 125~ per second have been tried,
two values being given to the potential difference at the terminals
of the alternator in each of the frequencies 72 and 125, making in
all 5 complete experiments. The curves so obtained are given in
Figs. 8, 9, 10, 11, and 12 respectively. From observations of the

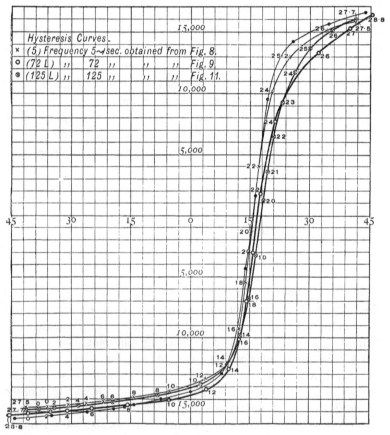

FIG. 5.

values of the electromotive force between C and D (Fig. 4) at
different times in the period, a curve A (in each experiment) was
plotted, giving the magnetising force in terms of the time; a

similar curve was plotted for the electromotive force between D and E, which, when corrected by subtracting the electromotive force due to the resistance of the coil H, gives the potential or time rate of variation of the induction in terms of the time. Hence the area of this curve (B) up to any point, *plus* a constant, is proportional to the induction corresponding to that point. This is shown in curve C, which is the integral of B. In each of the five experiments the ring with the non-inductive resistance was placed across the terminals of the alternator, and the excess of potential taken up by a non-inductive resistance.

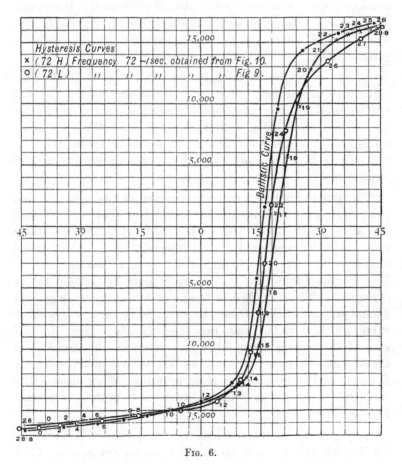

FIG. 6.

In Fig. 5 the hysteresis curves for frequencies of 5, 72, and 125 are compared with the ballistic curve. These curves are

marked 5, 72*L*, and 125*L* respectively. The corresponding values
for *B* and *H*, from which these curves have been plotted, are
given in Tables II., III., V., which have been obtained from the
curves in Figs. 8, 9, and 11 respectively.

The most noteworthy features in these curves are that the
curve with a frequency of 5 is very near the ballistic curve, if
allowance is made for difference in the magnetising current, and
that the curves with a frequency of 72 and 125 deviate very
materially, particularly in the part of the curve somewhat preceding the maximum induction. Hence the time effect mainly
develops with a greater frequency than 5 per second. Hence also
we infer that this effect, as already described in the *Electrician*,
is a true time effect, not arising in any way from the ballistic
galvanometer.

In Fig. 6 the hysteresis curves for a frequency of 72 are compared with the ballistic curve. The curves are marked 72*L* and
72*H* respectively, the potentials at the terminals of the alternator
in the two cases being approximately 36 and 430 volts. The
corresponding values for *B* and *H* are given in Tables III., IV.,
which have been obtained from the curves in Figs. 9 and 10
respectively.

The difference between the two curves in Fig. 6 was at first
puzzling, but a little consideration satisfied us that it arises from
the same time effect. The curve 72*L* was determined three
times, with the same result. The numerals refer to thirtieths of
a half-period. From 26 to 28·8 of the *L* curve the magnetising
force increases from 31·8 to 45·6, whilst from 21 to 26 of the *H*
curve it increases from 30·6 to 44, the rate of change being about
double as great in the former case as in the latter, and it is the *L*
curve which deviates most from the ballistic curve. In like
manner, in the neighbourhood of zero induction, the induction in
the *H* curve is changing twice as fast as the induction of the
L curve, and it is here the *H* curve which differs most. How
these differences of rate of change arise can be seen by inspecting
Figs. 9 and 10.

In Fig. 7 the hysteresis curves for a frequency of 125 are compared with the ballistic curve. The curves are marked 125*L*
and 125*H* respectively, the potentials at the terminals of the
alternator being approximately 62 and 750 volts. The cor-

responding values for B and H are given in Tables V. and VI., which have been obtained from the curves in Figs. 11 and 12 respectively.

These curves show the same difference as Fig. 6, but less markedly than in Fig. 5. The L curve was determined twice.

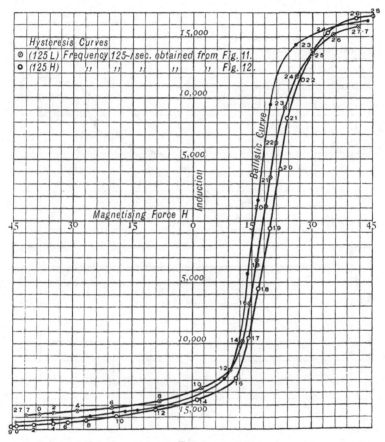

FIG. 7.

Experiments have been made upon chromium steel, supplied by Mr Hadfield, having the following composition:—0·71 per cent. carbon, 9·18 per cent. chromium, when annealed, and when hardened by raising to low yellow and plunging into cold water. The results show that the same time effect exists in this case, although it was not so marked as in the case of the hard stee

We draw the following conclusions from these experiments:—
(1) As Professor Ewing has already observed, after sudden change
of magnetising force, the induction does not at once attain to its
full value, but there is a slight increase going on for some seconds.
(2) The small difference between the ballistic curve of magnetisation with complete cycles and the curve determined with a con-

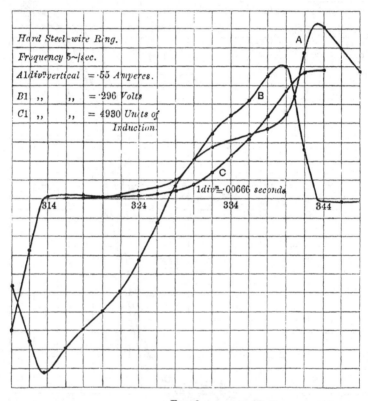

FIG. 8.

siderable frequency, which has already been observed, is a true
time effect, the difference being greater between a frequency of
72 per second and 5 per second than between 5 per second and
the ballistic curve.

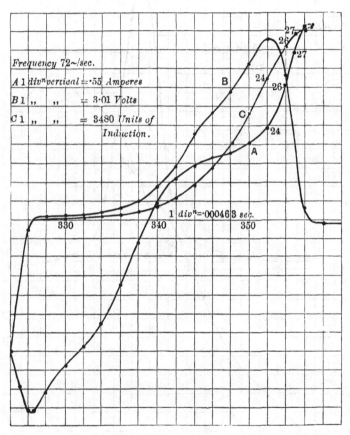

Frequency 72~/sec.

A 1 div^n vertical = ·55 Amperes
B 1 „   „   = 3·01 Volts
C 1 „   „   = 3480 Units of
              Induction.

1 div^n = ·00046 3 sec.

330   340   350

Fig. 9.

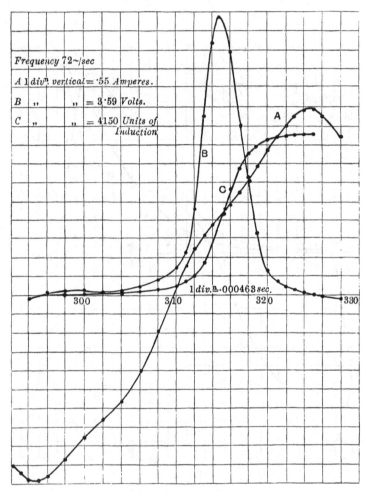

Frequency 72~/sec

A 1 div". vertical = ·55 Amperes.

B   „   „   = 3·59 Volts.

C   „   „   = 4150 Units of
                              Induction

1 div. = ·000463 sec.

300   310   320   330

FIG. 10.

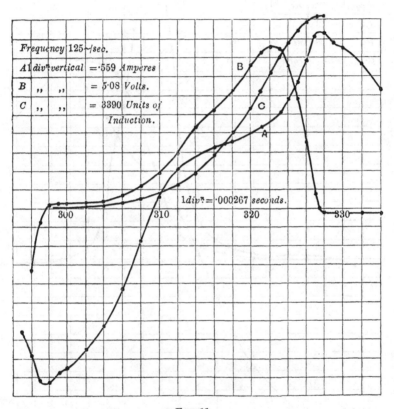

Fɪɢ. 11.

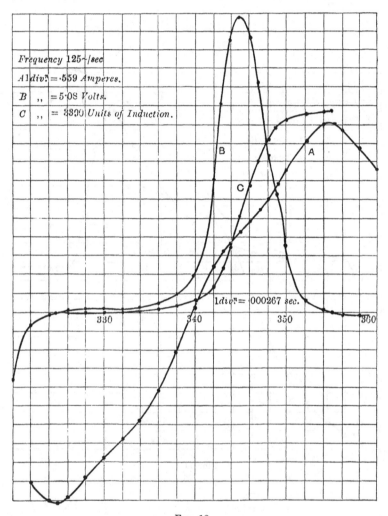

Fig. 12.

TABLE I.—*Hard Steel-wire Ring.*

H	B Points marked × obtained by method shown in Fig. 1	B Point marked • obtained by method shown in Fig. 2
+ 44·12	16,295	16,436
34·77	15,830	15,650
25·44	+ 14,407	14,290
19·32	..	9,539
16·1	..	+ 1,704
14·2	− 4,045	− 4,130
7·73	12,690	12,820
0	..	14,280
− 7·73	14,870	14,990
14·2	15,270	15,380
16·1	..	15,460
19·32	..	15,630
25·44	15,733	15,860
34·77	16,033	16,150
44·12	16,295	16,436

TABLE II.—*Frequency, 5 per second.*

B	H
15,660	41·7
15,200	34·4
13,010	25·1
10,190	19·95
+ 3,970	17·1
− 1,230	15·3
5,340	14·1
9,176	12·3
12,007	9·4
13,377	+ 3·4
14,382	− 5·6
14,747	14·4
15,203	22·0
15,295	26·85
15,477	30·86
15,523	35·75
15,660	41·7

TABLE III.—*Frequency, 72 per second. Potential at Terminals of Alternator, approximately 36 Volts.*

B	H
16,245	+ 45·7
16,180	45·5
15,215	39·9
13,410	31·7
7,805	21·3
+ 1,805	17·8
− 3,030	15·8
7,027	14·3
10,121	12·6
12,506	9·86
14,118	4·26
14,956	− 5·44
15,407	15·86
15,729	24·77
15,923	31·1
16,116	35·4
16,219	40·9
16,245	45·7

TABLE IV.—*Frequency, 72 per second. Potential at Terminals of Alternator, approximately 430 Volts.*

B	H
16,221	43·98
16,214	44·32
16,069	42·75
15,919	40·61
15,685	37·25
15,299	34·33
14,299	30·97
12,689	27·6
9,839	24·91
5,539	21·88
+ 999	19·3
− 5,073	16·6
9,609	14·36
12,300	11·22
13,530	7·18
14,145	+ 2·24
14,991	− 8·08
15,452	17·72
15,644	25·13
15,814	29·62
15,914	33·66
16,122	38·37
16,221	43·98

TABLE V.—*Frequency, 125 per second. Potential at Terminals of Alternator, approximately 62 Volts.*

B	H
15,936	+41·74
15,746	40·95
15,119	35·00
13,739	30·07
11,732	26·48
9,222	23·11
6,462	20·87
3,576	19·19
+ 1,192	18·18
− 3,136	16·16
6,776	14·59
9,850	12·57
12,172	9·20
13,615	2·24
14,618	− 8·08
15,120	19·75
15,434	28·61
15,622	34·10
15,773	38·37
15,936	41·74

TABLE VI.—*Frequency, 125 per second. Potential at Terminals of Alternator, approximately 750 Volts.*

B	H
16,689	+45·1
16,671	44·65
16,565	40·72
15,311	33·66
13,930	30·18
11,544	27·15
8,411	23·78
+ 4,077	21·54
− 565	19·30
5,396	16·83
9,474	14·36
12,862	10·77
14,368	1·12
15,309	− 9·20
15,873	18·85
16,099	26·25
16,262	30·63
16,413	34·33
16,564	39·49
16,670	43·98
16,689	45·10

37.

PROPAGATION OF MAGNETISATION OF IRON AS AFFECTED BY THE ELECTRIC CURRENTS IN THE IRON. By J. Hopkinson, F.R.S., and E. Wilson*.

[From the *Philosophical Transactions of the Royal Society of London*, Vol. CLXXXVI. (1895) A, pp. 93—121.]

Received May 17,—Read May 31, 1894.

PART I.

IT is not unfamiliar to those who have worked on large dynamos with the ballistic galvanometer, that the indications of the galvanometer do not give the whole changes which occur in the induction. Let the deflections of the galvanometer connected to an exploring coil be observed when the main current in the magnetic coils is reversed. The first elongation will be much greater than the second in the other direction, and probably the third greater than the second—showing that a continued current exists in one direction for a time comparable with the time of oscillation of the galvanometer. These effects cannot be got rid of, though they can be diminished by passing the exciting current through a non-inductive resistance and increasing the electro-motive force employed. This if carried far enough would be effective if the iron of the cores were divided so that no currents

* The experimental work of this paper was in part carried out by three of the Student Demonstrators of the Siemens Laboratory, King's College, London, Messrs Brazil, Atchison, and Greenham. We wish to express our thanks to them for their zealous co-operation.

could exist in the iron; but the currents in the iron, if the core is solid, continue for a considerable time and maintain the magnetism of the interior of the core in the direction it had before reversal of current. It was one of our objects to investigate this more closely by ascertaining the changes occurring at different depths in a core in terms of the time after reversal has been made.

The experiments were carried out in the Siemens Laboratory, King's College, London; and the electro-magnet used is shown in Fig. 1. It consists in its first form, the results of which though

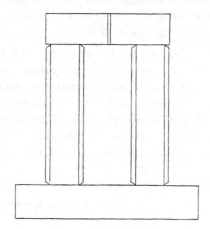

FIG. 1.

instructive are not satisfactory, of two vertical wrought-iron cores, 18 inches long and 4 inches diameter, wound with 2595 and 2613 turns respectively of No. 16 B.W.G. cotton-covered copper wire—

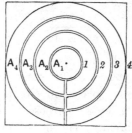

FIG. 2.

the resistance of the two coils in series being 16·3 ohms. The yoke is of wrought-iron 4 inches square in section and 2 feet long.

The pole-pieces are of wrought-iron 4 inches square, and all surfaces in contact are truly planed. One of the pole-pieces is turned down at the end, which butts on the other pole-piece, for half an inch of its length to a diameter of 4 inches; and three circular grooves are cut in the abutting face having mean diameters of 2·6, 5·16, and 7·75 centims. respectively, for the purpose of inserting copper coils the ends of which are brought out by means of the radial slot shown in Fig. 2. When the pole-pieces are brought into contact as shown in Fig. 1, we have thus three exploring coils within the mass and a fourth was wound on the circular portion outside. These exploring coils are numbered 1, 2, 3, 4 respectively, starting with the coil of least diameter.

Fig. 3 gives a diagram of the apparatus and connexions, in which A is a reversing switch for the purpose of reversing a current given by ten storage cells through the magnet windings in series; B is a Thomson graded galvanometer for measuring current; and C is a non-inductive resistance of about 16 ohms placed across the magnet coils for the purpose of diminishing the violence of the change on reversal. The maximum current given

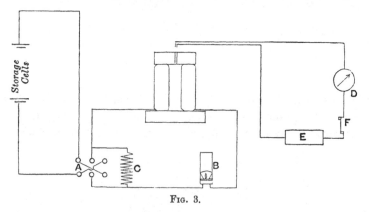

Fig. 3.

by the battery was 1·2 ampères. A D'Arsonval galvanometer of Professor Ayrton's type, D, of 320 ohms resistance; a resistance box E; and a key F were placed in circuit with any one of the exploring coils 1, 2, 3, 4, for the purpose of observing the electromotive force of that circuit. The method of experiment was as follows:—The current round the magnet limbs was suddenly reversed and readings on the D'Arsonval galvanometer were taken on each coil at known epochs after the reversal. The results are shown in Fig. 4, in which the ordinates are the electromotive forces in C.G.S. units and the abscissæ are in seconds.

The portion of these curves up to two seconds was obtained by means of a ballistic galvanometer having a periodic time of fifty seconds, the key of its circuit being broken at known epochs after reversal. From the induction curve so obtained the electromotive force was found by differentiation.

The curve *A* which is superposed on curve 4 of Fig. 4 gives the current round the magnet in the magnetising coils. It is worth noting, that, as would be expected, it agrees with the curve 4. The potential of the battery was 1·2 ampères × 16·3 ohms = 19·6 volts. Take the points two seconds after reversal, the electromotive force in one coil is 330,000; multiplying this by 5208, the number of coils on the magnet, we have in absolute

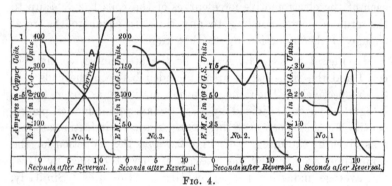

FIG. 4.

units 1,718,640,000 as the electromotive force on the coil due to electromagnetic change, or, say, 17·2 volts. Subtracting this from 19·6 we have 2·4. The electromotive force observed is

$$\cdot 125 \times 16\cdot 3 = 2\cdot 02.$$

The difference between these could be fully accounted for by an error of ¼ second in the time of either observation.

The general character of the results was quite unexpected by us. Take coil No. 2 for example, the spot of light, on reversing the current in the magnet winding, would at once spring off to a considerable deflection, the deflection would presently diminish, attaining a minimum after about 6 seconds; the deflection would then again increase and attain a maximum greater than the first after 8 seconds, it would then diminish and rapidly die away.

To attempt a thorough explanation of the peculiarities of these curves would mean solving the differential equation connecting induction with time and radius in the iron with the true relation of induction and magnetising force. But we may inversely from these curves attempt to obtain an approximation to the cyclic curve of induction of the iron.

Let l be the mean length of lines of force in the magnet. Let n be the number of convolutions on the magnet, and let c be the current in ampères in the magnetising coils at time t. Then at this epoch the force due to the magnetising coils is $4\pi nc/10l$. Call this H_1.

Next consider only one centimetre length of the magnet in the part between the pole-pieces which is circular and has coils 1, 2, 3, wound within its mass, and coil 4 wound outside. The area of each of the electromotive force curves of the coils 1, 2, 3, 4, up to the ordinate corresponding to any time, is equal to the total change of the induction up to that time.

In Fig. 2 let A_1, A_2, A_3, A_4 be the areas in sq. centims. of coil 1 and the ring-shaped areas included between the coils 1, 2, 3, 4 respectively. Then the induction at time t, as given by the integral of curve 1, divided by A_1 is the average induction per sq. centim. for this epoch over this area. Also, the induction at time t, as given by the integral of curve 2, *minus* the induction for the same time, as given by the integral of curve 1, divided by A_2, is the average induction per sq. centim. for this area. Similarly, average induction per sq. centim. for A_3, A_4 can be found for any epoch.

Consider area A_1. It is obvious that all currents induced within the mass considered *external* to this area, due to changes of induction, *plus* the current in the magnetising coil per centim. linear, at any epoch, go to magnetise this area, and, further, the induced currents in the outside of the area A_1 itself go to magnetise the interior portion of this area. We know the electromotive forces at the radii 1, 2, 3, 4, and the lengths in centims. of circles corresponding to these radii. From a knowledge of the specific resistance of the iron we can find the resistance, in ohms, of rings of the iron corresponding to these radii, having a cross-sectional area of 1 sq. centim. Let these resistances be respectively r_1, r_2, r_3, r_4. At time t, let e_1, e_2, e_3, e_4 be the electromotive forces in

volts at the radii 1, 2, 3, 4, then $\dfrac{e_1}{r_1}, \dfrac{e_2}{r_2}, \dfrac{e_3}{r_3}, \dfrac{e_4}{r_4}$ are at this epoch
the ampères per sq. centim. at these radii.   Let a curve be drawn
for this epoch, having ampères per sq. centim. for ordinates and
radii in centims. for abscissæ.   Then the area of this curve, from
radius 1 to radius 4, gives approximately the ampères per centim.
due to changes of induction, and (neglecting the currents within
the area considered) the algebraic sum of this force (call it H_2),
with the force due to the magnetising coils (H_1) at the epoch
chosen, gives the resultant magnetising force acting upon area A_1.
If H is this resultant force, we have $H = H_1 + H_2$.   Next draw a
curve showing the relation between the induction per sq. centim.
(B) and the resultant force (H) for different epochs.   This curve
should be an approximation to the cyclic curve of induction of the
iron.

The attempt to obtain an approximation to the cyclic curve of
induction from the curves in Fig. 4 was a failure, that is to say,
the resulting curve did not resemble a cyclic curve of magnetisation.   This is due to imperfections of fit of the two faces, in one
of which the exploring coils are imbedded.   That this imperfection
of fit will tend to have a serious effect upon the distribution of
induction over the whole area is obvious on consideration.   Take
the closed curve $abcd$ in Fig. 5, where AB is the junction between

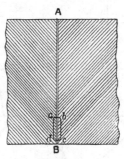

Fig. 5.

the pole-pieces.   If the space between the faces was appreciable,
the force along bc and ad in the iron could be neglected in comparison with the forces in the non-magnetic spaces ab, cd.   The
magnetising force is sensibly $4\pi c$, where c is the current passing
through the closed curve.   This may be made as small as we
please.   Therefore, the force along ab is equal to the force along

dc. In our case the space between the faces is very small, but has still a tendency towards an equalizing of the induction per unit area over the whole surface.

To test this the following experiment was tried. At a distance of $2\frac{1}{2}$ inches from the abutting surfaces of the pole-pieces four holes were drilled in one of the pole-pieces in a plane parallel with the abutting surfaces, as shown in Fig. 6. By means of a

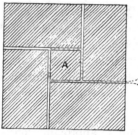

Fig. 6.

hooked wire we were able to thread an insulated copper wire through these holes, so as to enclose only the square area *A*, which is bounded by the drilled holes and has an area of ·61 sq. inch. The wire is indicated by the dotted lines. Fig. 7 gives

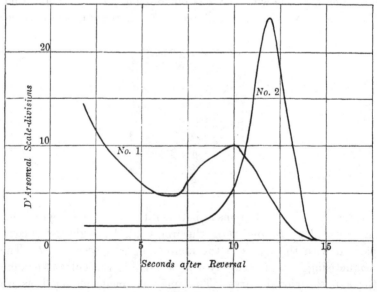

Fig. 7.

two curves taken by the D'Arsonval in the manner already described for a reversal of the same current in the copper coils of the magnets. No. 1 (Fig. 7) is the curve obtained from No. 1 coil (Fig. 2) near the air space. No. 2 (Fig. 7) is the curve obtained from the square coil shown in Fig. 6. The difference is very marked and shows at once the effect of the small non-magnetic space which accounts for the large initial change of induction previously observed on the coils 1, 2, 3 in Fig. 4. Similar holes were drilled in the yoke of the magnet in a plane midway between the vertical cores, having the same area of ·61 sq. inch; and on trial exactly the same form of curve was produced as is shown in No. 2 of Fig. 7. This method of drilling holes in the mass is open to the objection that the form of the area is square.

Whilst the above experiments were being made the portion of the magnet to take the place of the pole-pieces previously used was being constructed as follows:—In Fig. 8 the portion of the

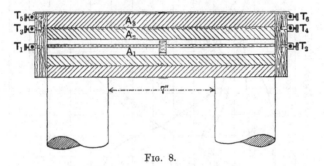

FIG. 8.

magnetic circuit resting upon the vertical cores consists of a centre rod A_1 of very soft Whitworth steel surrounded by tubes A_2, A_3 of the same material. The diameter of A_1 is 1 inch. The outside diameter of A_2 is $2\frac{1}{2}$ inches; and A_3 is 4 inches outside diameter between the cores of the magnet, but is 4 inches square at each end where it rests upon the magnet limbs. At the centre of the rod A_1 (longitudinally) a circular groove is turned down 1 millim. deep and 5 millims. wide, and also a longitudinal groove 1 millim. deep and 1 millim. wide is cut as shown in the figure for the purpose of leading a double silk-covered copper wire from terminal T_1 to 9 convolutions at the centre and along the rod to terminal T_2. A similar groove is cut in the outside of the tube A_2, and a copper wire is carried from terminal T_3 to 9 convolutions round the centre of the tube again along the groove to terminal T_4. Nine convolutions were also wound round the outside tube A_3, the ends of which are connected to the terminals T_5, T_6 respectively.

The tubes and rod were made by Sir J. Whitworth and Co., of Manchester, and a considerable force was required to drive the pieces into their proper position. Our best thanks are due to Professor Kennedy and his assistants for the putting together of these pieces by means of a 50-ton hydraulic testing machine. We are aware that the surfaces are somewhat scored by the hydraulic pressure, and the magnetic qualities may be slightly different for layers of the soft steel near these surfaces, but they serve just as well for the purpose of our experiments.

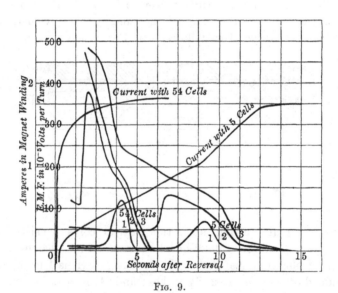

Fig. 9.

Systematic experiments were then commenced. The magnetising coils on the magnets were placed in parallel with one another, and a total current of 1·75 ampères (that is, ·87 ampère in each coil), due to 5 storage cells, was reversed through the coils. The arrangement of apparatus is shown in Fig. 3, except that the pole-pieces are replaced by the soft steel tubes shown in Fig. 8, and the non-inductive resistance C is removed. We have now three exploring coils instead of four, and these are marked

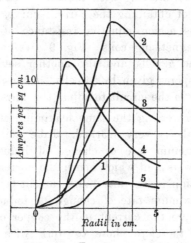

FIG. 9A.

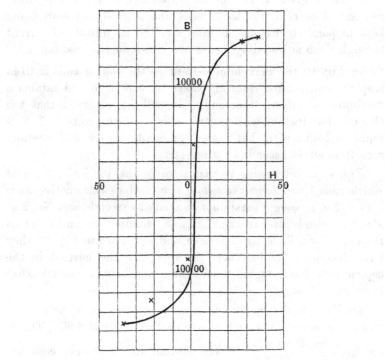

FIG. 9B.

1, 2, 3 respectively, starting with the coil of smallest diameter. For the purpose of obtaining the current curve, the D'Arsonval was placed across a non-inductive resistance of $\frac{1}{8}$ ohm in the circuit of the magnetising coils. Fig. 9 gives a set of curves obtained with the 5 cells, and also another set obtained by a reversal of 1·8 ampères given by 54 cells—a non-inductive resistance being placed in the circuit to adjust the current.

The effect of reversing the same maximum current with two different potentials is very marked. Take coil No. 1. With 5 cells the maximum rate of change of induction occurs at 9 seconds after reversal, at which epoch the current in the copper coils is about 1 ampère, the maximum current being 1·75. With 54 cells the maximum rate of change of induction occurs at 4 seconds, and here the current in the copper coils is nearly a maximum. We therefore chose to work with 54 cells, thus avoiding a magnetising force due to the current in the copper coils varying for considerable times after reversal.

Table I. gives a list of the experiments made with total reversal of current due to 54 cells, the magnetising coils being kept in parallel with one another, and the magnitude of current through them adjusted by means of a non-inductive resistance.

In Fig. 10 the maximum current in the copper coils is ·0745 ampère, which, after reversal, passes through zero and attains a maximum at about 3 seconds. It will be observed that the change of induction with regard to each of the coils 1, 2, 3 is rapid to begin with, but that it gradually decays and becomes zero at about 46 seconds after reversal.

Fig. 11 is interesting in that it gives the particular force at which coils 1 and 2 show a *second* rise in the electromotive force curves, No. 1 being a maximum at about 25 seconds, and No. 2 at about 8 seconds after reversal. These "humps" become a flat on the curve for a little smaller force, and, as shown in Fig. 10, they have disappeared altogether. In this case the current in the copper coils has attained a maximum at about 4 seconds after reversal.

In Fig. 12 the maximum current in the copper coils is ·24 ampère, corresponding with a force in C.G.S. units of 4·96. This is got from $\frac{4\pi}{10} \frac{2600 \times \cdot 24}{158}$. The current in the copper coils has

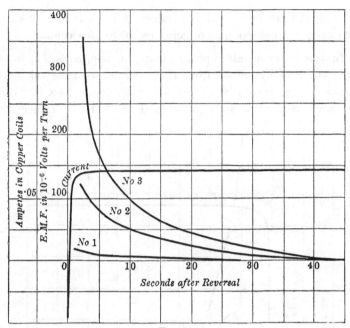

Fig. 10.

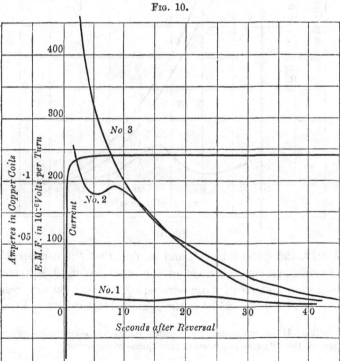

Fig. 11.

attained its maximum value at about 4 seconds after reversal, and changes of induction were going on up to 35 seconds.

In the following attempt to obtain an approximation to the cyclic curve of hysteresis, from these curves, we have taken the volume-specific resistance of the soft steel to be 13×10^{-6} ohm. We have taken the radii of coils 1, 2, 3 to be respectively

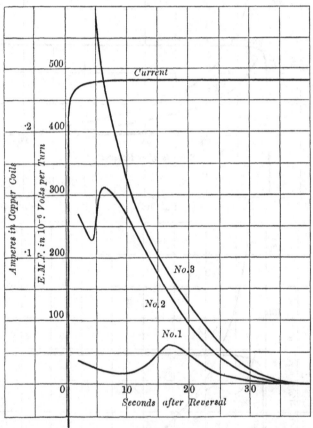

Fig. 12.

1·22, 3·18, and 5·08 centims.*, and we find that the corresponding resistances, in ohms, of rings of the steel having 1 sq. centim. cross-section and mean diameters equal to the coils are, respectively, $103·7 \times 10^{-6}$, $259·4 \times 10^{-6}$, and $416·4 \times 10^{-6}$. From a

* In Part II. of this paper the smallest radius was taken to be 1·27. For our purpose the difference is not worth the expense of correction.

knowledge of the electromotive forces at the three radii, for a
given epoch, we are able to find the ampères per sq. centim. at
those radii.   In Fig. 12 A a series of curves have been drawn for
different epochs, giving the relation between ampères per sq.
centim. and radii in centims., and the areas of these curves
between different limits have been found, and are tabulated in
Table II.   It is necessary here to state that the path of these
curves through the four given points in each case is assumed; we
have simply drawn a fair curve through the points.   But what we
wish to show is that the results obtained with the curves, drawn
as shown in Fig. 12 A, are not inconsistent with what we know
with great probability to be true.

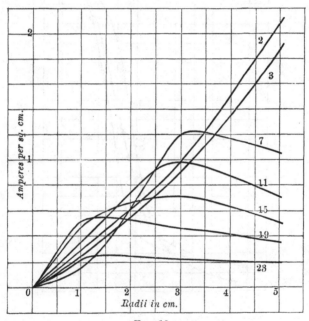

Fɪɢ. 12ᴀ.

The results shown in Fig. 12 B have been obtained as follows:
take curve I., Fig. 12 B; the electromotive force curve of coil 1,
Fig. 12, has been integrated, and the integral up to the ordinate
corresponding to any time is equal to the total change of the
induction up to that time, which divided by the area of the coil
in sq. centims. gives the average induction per sq. centim.   In
obtaining the areas we had to assume the path of the electro-

motive force curve up to 2 seconds, but this we can do with a good deal of certainty.

With regard to the forces we see that after 3 seconds the induced currents have to work against a constant current in the copper coils. In obtaining the forces due to induced currents we have only taken the area of the curves in Fig. 12 A between the radii 1·22 centims. and 5·08 centims.; that is, we have neglected the effect of the currents within the area of coil No. 1 altogether.

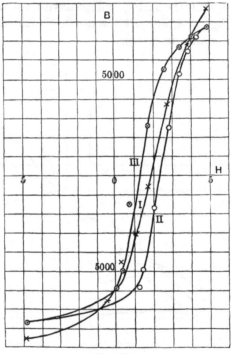

Fig. 12B.

The resultant force (H) is the algebraic sum of the force (H_2) due to the currents between the radii taken, and the force (H_1) due to the current in the copper coils, and is set forth for different epochs in Table II. The inductions per sq. centim. have been plotted in terms of this resultant force (H), and curve I., Fig. 12 B, shows this relation.

Next, take curves II. and III., Fig. 12 B. In obtaining the inductions for these curves, the difference between the integrals of curves No. 1 and 2, Fig. 12, for a given epoch, has been taken. This gives the induction for this epoch, which, when divided by the ring-shaped area between coils 1 and 2, gives the average induction per unit of that area.

In obtaining the forces in curve II., Fig. 12 B, we have taken the areas of the curves in Fig. 12 A between the radii 3·18 centims. and 5·08 centims.; that is, we have neglected the forces within the area under consideration as before. Here the error is of more importance, and may partly account for the difference between the forces of curves I., II. In curve III. we have taken the areas of curves in Fig. 12 A between the radii 2·2 and 5·08; that is, we

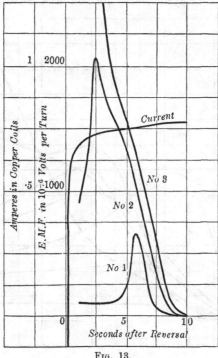

FIG. 13.

have taken account of the force due to induced currents over a considerable portion of the area considered. Coupled with the uncertainty in form of the curves in Fig. 12 A we have the uncertainty as to how much to allow for the forces due to induced currents over the particular area considered. The difference in the ordinates of curves I. and II. may partly be accounted for by

errors arising from the assumed path of the electromotive force
curve up to 2 seconds, which is more uncertain in curve 2, Fig. 12,
than in curve 1 ; and partly to possible slight inequality between
the materials of the rod and its surrounding tube.

In Fig. 13 the maximum current in the copper coils is ·77
ampère, corresponding with a force in C.G.S. units of 16. The
current in the copper coils, after passing through zero, attains its
full value at about 9 seconds after reversal, and the change of
induction ceases at 10 seconds.

No. 1 curve, Fig. 13, has been integrated, and the maximum
induction per sq. centim. found to be 14,500 C.G.S. units. We
have taken a given cyclic curve for soft iron corresponding with

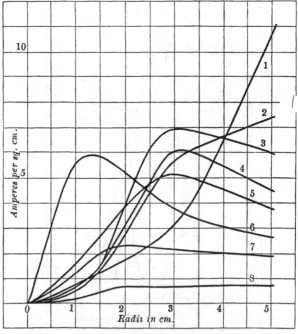

Fig. 13A.

this maximum induction, and have tabulated the forces obtained
therefrom in Table III. for the different values of B got from the
integration of No. 1 curve. We then plotted in Fig. 13 A the
ampères per sq. centim. at the different radii for different epochs,
and in each case, by drawing a curve fairly through them, we
were able to produce areas in fair correspondence with areas as

got by means of the given cyclic curve. The comparative areas are tabulated in Table III.

In Fig. 9 the maximum current in the copper coils due to the 54 cells is 1·8 ampères, corresponding with a force of 20·7 in C.G.S. units. In this case the current had passed through zero and attained a maximum at 6 seconds after reversal; the change of induction being zero also at this time. We have worked out the current per sq. centim. for the different radii at different epochs, as before, and have plotted them in Fig. 9 A. Fig. 9 B gives the relation of B to H, found from the curves, and it also shows a fair approximation to the cyclic curve for soft iron, although in this case the points are fewer in number and were more difficult to obtain, owing to the greater rapidity with which the D'Arsonval needle moved as compared with the earlier curves.

With a reversal of 2·3 ampères the whole induction effects had died out at 5 seconds after reversal. Coil No. 1 showed a maximum electromotive force at about $3\frac{1}{2}$ seconds. Coil No. 2 gave a dwell, and attained a maximum at 2 seconds, and then died rapidly away. Coil No. 3 attained an immediate maximum and died rapidly to zero at 5 seconds.

With a reversal of $6\frac{1}{2}$ ampères the whole inductive effects had died out at about 3 seconds after reversal. No. 1 coil showed a maximum electromotive force at about $1\frac{3}{4}$ seconds. No. 2 gave a dwell and attained a maximum at about $1\frac{1}{2}$ seconds and rapidly died away to zero at about 2 seconds. No. 3 attained an immediate maximum and died rapidly to zero at about 2 seconds.

The variations in form of these curves and of the times the electromotive forces take to die away are intimately connected with the curve of magnetisation of the material. When the magnetising force is small (1·7) the maxima occur early because the ratio induction to magnetising force is small. As the magnetising force increases to 3 and 4·96 the maxima occur later because this ratio has increased, whilst when the force is further increased to 16 and 37·2, as shown in Figs. 13 and 9, the maxima occur earlier because the ratio has again diminished.

The results, both of these experiments and of those which follow, have a more general application than to bars of the particular size used. From the dimensions of the partial differential equation which expresses the propagation of induction in the bar, one sees at once that if the external magnetising forces are the same in two bars differing in diameter, then similar magnetic events will occur in the two bars, but at times varying as the square of the diameters of the bars. But one may see this equally without referring to the differential equation. Suppose two bars, one n times the diameter of the other, in which there are equal variations of the magnetising forces; consider the annulus between radii r_1, r_2 and nr_1, nr_2 in the two, the resistance per centimetre length of the rods of these annuli will be the same for their area, and their lengths are alike as $1 : n$; the inductions through them, when the inductions per centimetre are the same, are as the areas, that is, as $1 : n^2$. Hence if the inductions change at rates inversely proportional to $1 : n^2$, the currents between corresponding radii will be the same at times in the ratio of $1 : n^2$, and the magnetising forces will also be the same.

Magnets of sixteen inches diameter are not uncommon; with such a magnet, the magnetising force being 37 and the magnetising current being compelled to at once attain its full value, it will take over a minute for the centre of the iron to attain its full inductive value.

On the other hand, with a wire or bundle of wires, each 1 millim. diameter, and a magnetising force between 3 and 5, which gives the longest times with our bar, the centre of the wire will be experiencing its greatest range of change in about $\frac{1}{500}$ second. This is a magnetising force similar to those used in transformers, and naturally leads us to the second part of our experiments.

PART II.—ALTERNATE CURRENTS.

This part of the subject has a practical bearing in the case of alternate current transformer cores, and the armature cores of dynamo-electric machines.

The alternate currents used have periodic times, varying from 4 to 80 seconds, and were obtained from a battery of 54 storage cells by means of a liquid reverser*, shown in elevation and plan

---

* This form of reverser is due to Professor Ewing.

in Figs. 14 and 15. It consists of two upright curved plates of sheet copper, *AA*, between which were rotated two similar plates, *BB*, connected with collecting rings, *DD*, from which the current was led away by brushes to the primary circuit of the magnet. The copper plates are placed in a weak solution of copper sulphate in a porcelain jar. The inner copper plates, and the collecting

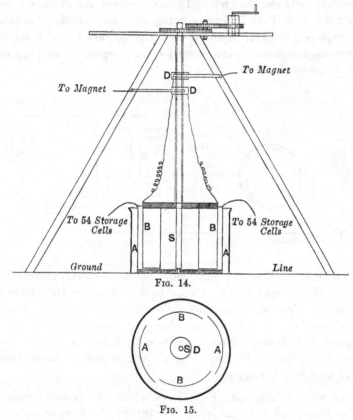

To Magnet

To Magnet

D

D

To 54 Storage Cells

B

S

B

To 54 Storage Cells

A

A

Ground

Line

Fig. 14.

B

A

S

D

A

B

Fig. 15.

rings, are fixed to a vertical shaft, *S*, which can be rotated at any desired speed by means of the gearing shown in the figure. The outer plates are connected to the terminals of the battery of storage cells, and the arrangement gives approximately a sine curve of current when working through a non-inductive resistance.

The experiments were made with the same electro-magnet and Whitworth steel tubes described in Part I. of this paper.

Fig. 16 gives a diagram of connexions in which M is the current reverser, G is the Thomson graded current meter for measuring the maximum current in the copper coils, and W is the electromagnet. A small, non-inductive resistance, placed in the primary circuit served to give the curve of current by observations on the D'Arsonval galvanometer, D, of the time variation of the potential difference between its ends. The D'Arsonval galvanometer was also used, as in Part I., for observing the electromotive forces of the exploring coils 1, 2, and 3 (see Fig. 8, Part I.), R being an adjustable resistance in its circuit for the purpose of keeping the deflections on the scale.

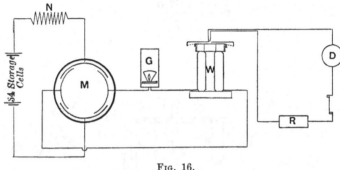

FIG. 16.

The method of experiment was as follows :—The liquid reverser, M, was placed so as to give a maximum current on the meter G, which was adjusted by non-inductive resistance, N, to the desired value, and, in all cases, when changing from higher to lower currents, a system of demagnetisation by reversals was adopted. Time was taken, as in Part I., on a clock beating seconds, which could be heard distinctly.

As an example, take Fig. 17, in which the periodic time is 80 seconds, and the maximum current in the copper coils ·23 ampère. The E.M.F. curves of the exploring coils are numbered 1, 2, and 3 respectively, and the curve of current in the copper coils is also given.

As in the case of simple reversals (Part I.) we may from these curves attempt to obtain an approximation to the cyclic curve of induction of the iron. In all cases where this is done we have taken coil 1 and considered the area within it—that is to say, from a knowledge of the E.M.F.'s at different depths of the iron, due to change of induction at any epoch, we have estimated the average magnetising force acting in this area, and this we call H_2. The curves from which these forces have been obtained are given in Fig. 17 A, and have been plotted from Table VI. The algebraic sum of this force, H_2, and the force H_1, given at the same epoch by the current in the copper coils, is taken to be the *then* resultant force magnetising this area.   Also the integral of curve 1, Fig. 17, gives the average induction over this area at the same epoch. Curve x, Fig. 17 B, is the cyclic curve obtained by plotting the inductions in terms of the resultant force H.

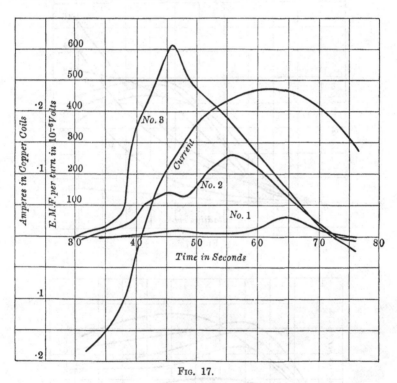

FIG. 17.

A word is necessary with regard to the last column in Table VI. This gives the total dissipation of energy by induced currents in ergs per cycle per cub. centim. of the iron. We know the watts per sq. centim. at different depths of the iron for different epochs. Let a series of curves be drawn (Fig. 17 c) for chosen epochs giving this relation: the areas of these curves from radii 0 to 5·08 give for the respective epochs the watts per centim. dissipated by induced currents.   In symbols this is $\int \dfrac{ec}{\text{sq. centim.}}\, dr$; where r is the radius, and e, c the E.M.F. and current.   It is now only necessary to integrate with regard to time in order to obtain the total dissipation: we have chosen a half period as our limits.   This gives us $\iint \dfrac{ec}{\text{sq. centim.}}\, dr\, dt$, and is got from the area of curve z, Fig. 17 D. The ordinates of this curve are taken from the last column of Table VI*.

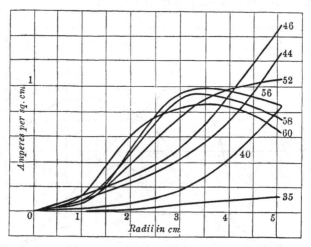

Fig. 17A.

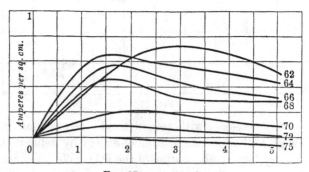

Fig. 17A.—continued.

* Figures 18c, 18D, 19c, 19D, 20c, and 20D have been omitted as they can readily be reconstructed from the tables and do not illustrate any general conclusion. [ED.]

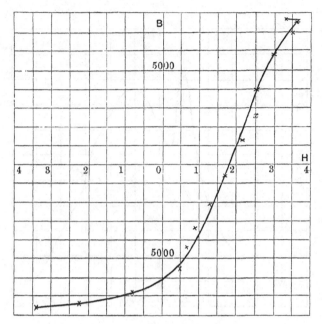

FIG. 17B.

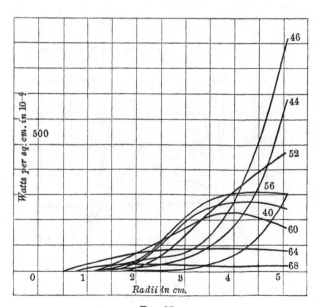

FIG. 17C.

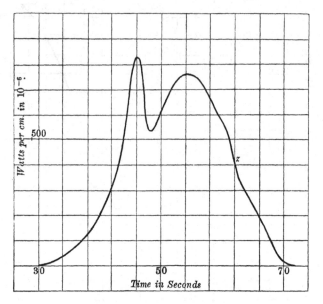

Fig. 17d.

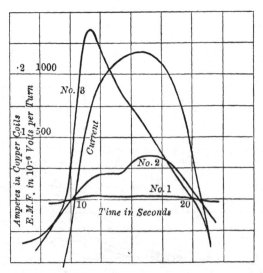

Fig. 18.

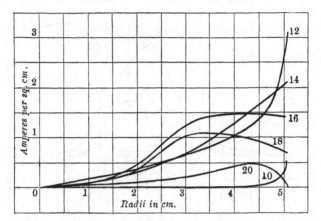

Fig. 18a.

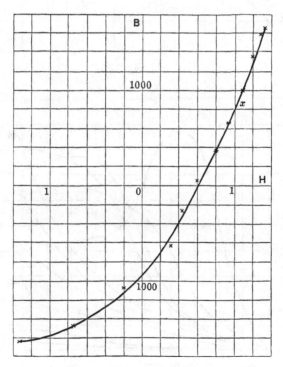

Fig. 18b.

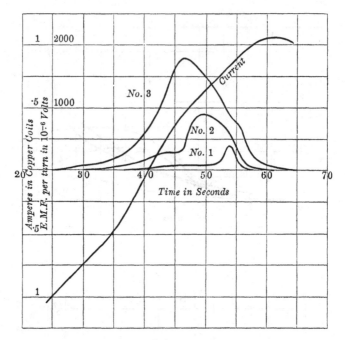

FIG. 19.

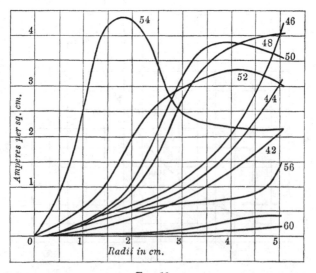

FIG. 19A.

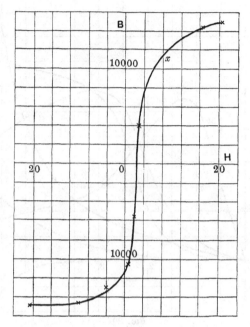

FIG. 19B.

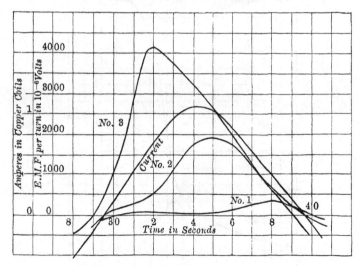

FIG. 20.

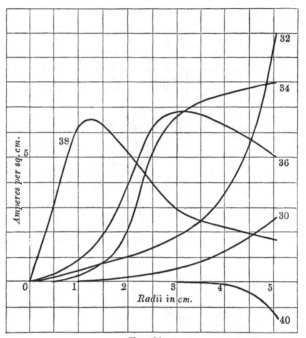

Fɪɢ. 20ᴀ.

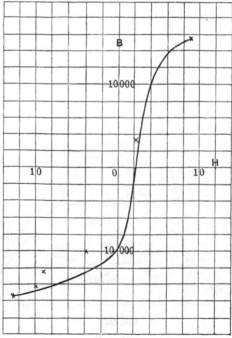

Fɪɢ. 20ʙ.

The curves in Figs. 19, 20 have been treated in a similar manner to that already described in connexion with Fig. 17. But in Fig. 18 the procedure is a little different. In this case the periodic time is 20, and the maximum force per centim. linear, due to the current in the copper coils, is 4·87. With this frequency and current the effects of induced currents in the iron are very marked: we have taken a given soft iron cyclic curve, of roughly the same maximum induction as given by the integral of curve No. 1, Fig. 18, and have tabulated the forces obtained therefrom in Table VII. In Fig. 18 A we have plotted the ampères per sq. centim. at the different radii, and for the several epochs, and in each case, by drawing a curve fairly through these points, as shown in the figure, we are able to produce areas in fair correspondence with the areas obtained by means of the given cyclic curve. The comparative areas are given in Table VII.

The results shown in Fig. 20 are by no means so satisfactory as the results given by other figures, but we have thought it better to insert them here, as we do not wish to make any selection of results which might give an idea of average accuracy greater than these experiments are entitled to.

Referring now to the summary of results in Table V., we note the marked effect of change of frequency upon the average induction per unit area of the innermost coil No. 1, when dealing with comparatively small maximum inductions. Compare the results given in Figs. 17 and 18. The maximum force per centim. linear due to the current in the copper coils is 4·8 in each case, but the average induction per sq. centim. of coil No. 1 is reduced from 7690 to 1630 by a change of frequency from $\frac{1}{80}$ to $\frac{1}{20}$. This is, of course, not the case on the higher portion of the induction curve, as is shown by the results of Figs. 19 and 20, although the resultant force H is reduced by the induced currents.

In Fig. 21 the maximum ampères in the copper coils is ·24, and the periodic time is reduced to 4. An inspection of these curves shows the marked effect of change of frequency, coil No. 2 being exceedingly diminished in amplitude as compared with No. 3.

As an example of the practical bearing of this portion of the paper, suppose we have a transformer core made out of iron wire, 1 millim. in diameter, the wires being perfectly insulated from

one another. The outside diameter of our outer tube is 101·6
millims. Similar events will therefore happen at times, varying
as $\left(\dfrac{1}{101·6}\right)^2$. Take the case of Fig. 17, in which the periodic time
is 80 seconds, and the maximum average induction per sq. centim.
is about 7000.

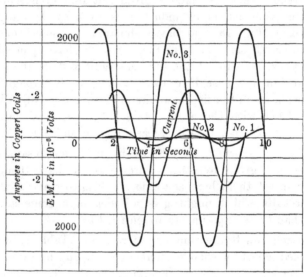

Fig. 21.

$\dfrac{(101·6)^2}{80} = 129$ periods per second, and this is an example which
might arise in practice. The ergs dissipated per cycle per cub.
centim. are 3820 by induced currents, and about 3000 by magnetic
hysteresis. We see further, from Fig. 18, that at 500 periods per
second only the outside layers of our 1 millim. wire are really
useful.

As another example, take the case of an armature core of a
dynamo-electric machine in which a frequency of 1000 complete
periods per minute might be taken.

In Fig. 19 the periodic time is 80, and the maximum average
induction per sq. centim. is 15,000.

We have
$$\tfrac{1}{17} = 80\,(x/101·6)^2;$$
$$x = 101·6/36 = \text{nearly 3 millims.}$$

The ergs dissipated per cycle per cub. centim. are 26,000 by induced currents, and about 17,000 by magnetic hysteresis. This shows that according to good practice, where the wires in armature cores are of an order of 1 or 2 millims. diameter, the loss by induced currents would be but small as compared with the loss by magnetic hysteresis. This, of course, assumes the wires to be perfectly insulated from one another, which is not always realised in practice.

Both the armature cores of dynamos and the cores of transformers are now usually made of plates instead of wire; roughly speaking a plate in regard to induced currents in its substance is comparable to a wire of a diameter double the thickness of the plate. We infer that the ordinary practice of making transformer plates about ½ millim. thick, and plates of armature cores 1 millim. thick, is not far wrong. Not much is lost by local currents in the iron, and the plates could not be much thicker without loss *.

TABLE I.

	Maximum ampères in magnetising coils	Maximum force in c.g.s. units H	Maximum induction per sq. centim. B
Fig. 10	·0745	1·7	
,, 11	·138	3·0	
,, 12, Table II.	·24	4·96	8,000
	·49	10·1	12,820
,, 13, ,, III.	·774	16·0	14,495
,, 9, ,, IV.	1·80	37·2	15,480
	2·31	47·6	
	6·5	134·5	

---

* The question of dissipation of energy by local currents in iron has been discussed by Professors J. J. Thomson and Ewing. See the *Electrician*, April 8th and 15th, 1892.

TABLE II.

Time in seconds after reversal	Curve I.			Curves II. and III.				Force due to current in magnet coils. H_1	Radius 1·22 cm.		Radius 3·18 cm.	
	Area in $\frac{4}{25}$ diagram squares of Curve 1. Fig. 12	Change of induction per sq. cm. in c.g.s. units	Induction per sq. cm. B	Area in $\frac{4}{25}$ diagram squares of Curve 2. Fig. 12	Area of Curve 2, minus Curve 1, in $\frac{4}{25}$ diagram squares	Change of induction per sq. cm. in c.g.s. units	Induction per sq. cm. B		E.M.F. in volts in 10^{-6}	Ampères per sq. cm.	E.M.F. in volts in 10^{-6}	Ampères per sq. cm.
0	0	0	− 8,680	0	0	0	− 7,720	− 4·96	0	0	0	0
2	1·5	..	..	13·7	12·2	1,842	− 5,878	+ 4·86	36	·377	268	1·06
3	2·7	2,130	− 6,550	20·9	18·2	2,744	− 4,976	+ 4·96	32	·335	247	·985
7	5·1	4,024	− 4,656	47·1	42	6,315	− 1,405	+ 4·96	20	·209	310	1·23
11	7·1	5,604	− 3,076	76·0	68·9	10,360	+ 2,460	+ 4·96	22	·230	247	·985
15	10·2	8,046	− 634	98·0	87·8	13,203	+ 5,483	+ 4·96	52	·544	180	·717
19	16	12,623	+ 3,943	111·5	95·5	14,360	+ 6,640	+ 4·96	52	·544	113	·450
23	19·7	15,540	+ 6,860	119·5	99·8	15,006	+ 7,286	+ 4·96	24	·251	52	·207
35	22	17,360	+ 8,680	124·7	102·7	15,440	+ 7,720	+ 4·96	0	0	0	0

Time in seconds after reversal	Radius 5·08 cm.		Curve I.			Curve II.			Curve III.		
	E.M.F. in volts in 10^{-6}	Ampères per sq. cm.	Area in 2 diagram squares. Radii 1·22 to 5·08 of Curves in Fig. 12 A	Force due to induced currents. H_2	Resultant force. H	Area in 2 diagram squares. Radii 3·18 to 5·08 of Curves in Fig. 12A	Force due to induced currents. H_2	Resultant force. H	Area in 2 diagram squares. Radii 2·2 to 5·08 of Curves in Fig. 12A	Force due to induced currents. H_2	Resultant force. H
0	0	0	0	− 0	− 4·96	0	0	− 4·96	0	0	− 4·96
2	888	2·14	21·4	− 5·3	− ·44	15·0	− 3·77	+ 1·09	19·1	− 4·8	+ ·06
3	787	1·89	21·1	− 5·28	− ·32	14·4	− 3·62	+ 1·34	18·4	− 4·6	+ ·36
7	437	1·05	18·9	− 4·73	+ ·23	11·6	− 2·92	+ 2·04	17·0	− 4·2	+ ·76
11	291	·699	14·6	− 3·6	+ 1·16	8·4	− 2·11	+ 2·85	12·7	− 3·2	+ 1·76
15	204	·490	12·9	− 3·2	+ 1·76	6·0	− 1·51	+ 3·45	9·7	− 2·4	+ 2·56
19	144	·346	8·56	− 2·14	+ 2·82	4·0	− 1·0	+ 3·96	6·4	− 1·6	+ 3·36
23	88	·211	4·23	− 1·05	+ 3·91	2·0	− ·50	+ 4·46	3·0	− ·75	+ 4·21
35	0	0	0	0	+ 4·96	0	0	+ 4·96	0	0	+ 4·96

TABLE III.

Time in seconds after reversal	Area in 4/7 diagram squares of Curve I, Fig. 13	Change of induction per sq. cm. in c.g.s. units	Induction per sq. cm. B	Radius 1·22 cm. E.M.F. in volts in 10^{-6}	Radius 1·22 cm. Ampères per sq. cm.	Radius 3·18 cm. E.M.F. in volts in 10^{-6}	Radius 3·18 cm. Ampères per sq. cm.	Radius 5·08 cm. E.M.F. in volts in 10^{-6}	Radius 5·08 cm. Ampères per sq. cm.	Force due to current in magnet winding, H_1	Force taken from the given cyclic curve, H	Difference of forces, H_1 and H	Area in 4 diagram squares corresponding to differences of forces, H_1 and H	Area in 4 diagram squares of curves in Fig. 13A
0	0	0	− 14,495	0	0	0	0	0	0	+ 16·02	16·0	0	0	0
1	0·92	1,814	− 12,680	99	·95	902	3·48	4615	11·1	− 14·09	3·0	17·0	6·7	7·9
2	1·9	3,747	− 10,748	91·3	·88	1534	5·91	3077	7·4	− 14·83	0	14·8	5·9	7·8
3	2·5	4,930	− 9,565	79·9	·77	1804	6·95	2460	5·9	− 15·13	0·5	14·6	5·8	8·0
4	3·7	7,298	− 6,197	91·3	·88	1579	6·09	1850	4·4	− 15·28	1·0	14·3	5·7	8·8
5	4·8	9,467	− 5,028	205	1·98	1331	5·13	1540	3·7	− 15·43	1·2	14·2	5·6	7·8
6	10·0	19,723	+ 5,228	647	5·87	902	3·48	1080	2·6	− 15·58	2·7	12·9	5·2	7·1
7	13·6	26,825	+ 12,330	152	1·47	541	2·09	810	1·9	− 15·8	8·0	7·8	3·1	3·9
8	14·4	28,400	+ 13,905	26·6	·26	158	·61	310	0·7	− 15·91	11·9	4·01	1·6	1·6
9	14·6	28,795	+ 14,300	3·8	·04	22·6	·09	77	0·2	− 16·02	14·3	1·72	·7	·8
10	14·7	28,990	+ 14,495	0	0	0	0	0	0	− 16·02	16·0	0	0	0

TABLE IV.

Time in seconds after reversal	Area in 3/5 diagram squares of Curve 1, Fig. 9	Change of induction per sq. cm. in c.g.s. units	Induction per sq. cm. B	Radius 1·22 cm. Ampères per sq. cm.	Radius 1·22 cm. E.M.F. in volts in 10^{-6}	Radius 3·18 cm. Ampères per sq. cm.	Radius 3·18 cm. E.M.F. in volts in 10^{-6}	Radius 5·08 cm. Ampères per sq. cm.	Radius 5·08 cm. E.M.F. in volts in 10^{-6}	Area in diagram squares of curves in Fig. 9a, radii 1·22 to 5·08	Force due to induced currents H_2	Force due to current in magnet winding H_1	Resultant force H
0	0	0	-15,480	0	0	0	0	0	0		0	-37·2	-37·2
1	·9	1,775	-13,705	1·17	116	4·58	1150	11·05	4595	22·2	-55·7	+34·1	-21·6
2	2·2	4,340	-11,140	1·17	116	15·0	3775	6·9	2872	14·2	-35·6	+34·1	-1·5
3	3·3	6,510	-8,970	1·52	145	9·2	2310	3·45	1436	14·2	-35·6	+36·4	+·8
4	9·8	19,300	+3,820	11·7	1160	6·25	1569	1·72	718	3·2	-8·0	+36·4	+28·4
5	15·6	30,800	+15,320	0	0	2·09	524	0	0	0	0	+37·2	+37·2
6	15·7	30,960	+15,480	0	0	0	0	0	0	0			

TABLE V.

Figures	Tables	Periodic time in seconds	Force due to current in copper coils		Max. resultant force. H	Max. average induction per sq. cm. of Coil 1, in c.g.s. units. B	Comparative areas of Curves Nos. 1, 2, 3, Figs. 17, 18, 19, 20			$\frac{1}{4\pi}\int HdB$ taken from Ewing	Ergs per cycle per cu. cm. dissipated by induced currents obtained from Figs. 17D, 18D, 19D, 20D
			Max. ampères	Max. H_1			No. 1	No. 2	No. 3		
17, 17A, 17B, 17C, 17D,	VI.	80	·23	4·87	3·67	7,690	9·7	62·5	139	3,000	3,820
18, 18A, 18B, 18C, 18D,	VII.	20	·23	4·87	1·32	1,630	1·5	22·7	69·4		2,750
		80	·35			11,300					
		20	·35			2,700					
		80	·74			16,000					
		20	·74			13,700					
19, 19A, 19B, 19C, 19D,	VIII.	80	1·06	21·9	21·8	15,000	3·3	22·2	60·0	17,000	26,200
20, 20A, 20B, 20C, 20D,	IX.	20	1·06	21·9	13·0	15,500	2·1	18·5	43·4		54,500
21		4	·24								

20—2

TABLE VI.—Periodic Time, 80 seconds. Maximum Current in Copper Coils, ·23 ampère.

Time in seconds	Coil 1				Coil 2				Coil 3			
	Deflection on D'Arsonval	E.M.F. in 10^{-6} volts per turn	Ampères per sq. cm.	Watts per sq. cm. in 10^{-6}	Deflection on D'Arsonval	E.M.F. in 10^{-6} volts per turn	Ampères per sq. cm.	Watts per sq. cm. in 10^{-6}	Deflection on D'Arsonval	E.M.F. in 10^{-6} volts per turn	Ampères per sq. cm.	Watts per sq. cm. in 10^{-6}
30	+ 0	..	..	..	+ 1	..	..	..	0	..	..	..
32	+ 0	..	..	..	0	..	..	..	− 1	17·5	..	..
34	+ 0	..	0	..	− 1	10·9	·06	..	− 1·5	26·3	·09	..
36	− 0	2·13	..	..	− 2	21·8	·08	1·74	− 3	52·6	·13	6·84
38	− ·5	4·26	·04	·17	− 3	32·7	..	..	− 5	87·7	..	..
40	− 1·0	8·51	..	..	− 5	54·5	·21	11·4	− 20	351	·85	298
42	− 2	14·9	·14	2·09	− 10	109	..	..	− 25	439	..	..
44	− 3·5	19·2	·18	3·46	− 11	120	·46	55·2	− 30	527	1·27	669
46	− 4·5	14·9	·14	2·09	− 13	142	·55	78·1	− 35	614	1·49	915
48	− 3·5	12·8	..	..	− 11·5	125	·48	60	− 30	527	1·27	669
50	− 3	12·8	·12	1·54	− 14	153	..	..	− 25	439	..	..
52	− 3	12·8	..	..	− 19	207	·80	165	− 25	439	1·06	465
54	− 3	12·8	·12	1·54	− 22	240	..	..	− 23	404	..	..
56	− 3	17·0	·16	2·72	− 23	251	·97	243	− 20	351	·85	298
58	− 4	25·5	·25	6·37	− 22	240	·93	223	− 18	316	·76	240
60	− 6	42·6	·41	17·5	− 20	218	·84	183	− 15	263	·63	166
62	− 10	63·8	·62	39·6	− 17	185	·71	131	− 12	211	·51	108
64	− 15	55·3	·53	29·3	− 13	142	·55	78	− 10	175	·42	73·5
66	− 13	42·6	·41	17·5	− 10	109	·42	45·8	− 7	123	·3	36·9
68	− 10	17·0	·16	2·72	− 7	76·3	·29	22·1	− 4	70·2	·17	11·9
70	− 4	8·5	·08	·68	− 4	43·6	·17	7·4	− 2	35·1	·08	2·8
72	− 2	2·13	·02	·04	− 1	10·9	·04	·45	0	0	0	0
74	− ·5	..	0	..	+ ·5	5·4	·02	·11	1	17·5	·04	·7
76	+ ·5	2·13	·02	·04	+ 1	10·9	·03	..	2·5	43·9	·7	..
78	+ 1·0	4·26	..	..	+ 2	21·8	·04	·45	5	87·7	·81	..

TABLE VI.—continued.

Time in seconds	Average induction per sq. cm. of Coil 1			Current in copper coils			Magnetising force due to induced currents		Resultant magnetising force per cm. linear. H_3	Energy dissipated by induced currents	
	Area of E.M.F. curve in $\frac{1}{25}$ diagram square	Half max. area ± other areas in sq. cm.	Induction per sq. cm. in c.g.s. units. B	Deflection on D'Arsonval	Ampères	Force per cm. linear. H_1	Area in diagram squares of curves in Fig. 17A, from rad. 1·27 to 5·08	Force per cm. linear. H_2		Area in diagram squares of curves in Fig. 17c, from rad. 0 to 5·08	Watts per cm. in 10^{-6}
30	··	··	··	+67	+·204						
32	··	··	··	+61	+·186						
34	0	+19·5	7690	+53	+·162	+3·35	2·5	·31	+3·41		
36	··	··	··	+45	+·137	+3·1					
38	·5	··	··	+34	+·104	+2·85					
40	··	+19·0	7496	+10	+·03	+·63	12·5	1·57	+2·2	3·9	195
42	2·15			-10	-·03						
44	··	+17·35	6844	-30	-·091	-1·90	22	2·77	+·87	10·5	525
46	5·8			-40	-·122					16·4	820
48	··	+13·7	5405	-50	-·152	-3·16	21	2·64	-·52	10·5	525
50	8·2			-59	-·18						
52	··	+11·3	4460	-66	-·20	-4·17	27·7	3·48	-·69	14·4	720
54	10·9			-70	-·21						
56	12·2	+8·6	3390	-73	-·22	-4·62	29	3·65	-·97	15	750
58	14	+7·3	2880	-75	-·229	-4·74	27·7	3·48	-1·26	12·8	640
60	17·7	+5·5	2170	-77	-·235	-4·87	28	3·52	-1·35	11·5	575
62	23	+1·8	710	-77	-·235	-4·87	24·9	3·13	-1·74	7·5	375
64	29·4	-3·5	1380	-77	-·235	-4·87	22	2·77	-2·10	5·8	290
66	34·1	-9·9	3910	-75	-·229	-4·74	17·5	2·20	-2·54		
68	37·3	-14·6	5760	-73	-·22	-4·62	13	1·63	-2·99		
70	38·5	-17·8	7020	-68	-·207	-4·30	6	·75	-3·55	1·5	75
72	··	-19·0	7500	-63	-·192	-3·98	2·5	·31	-3·67		
74	39·1	··	··	-55	-·168	-3·48					
76	··	-19·5	7690	-47	··	-3·2	·6	·07	-3·3		
78	··	··	··	-35	-·143	-2·97					

TABLE VII.—Periodic Time, 20 seconds.   Maximum Current in Copper Coils, ·23 ampère.

Time in seconds	Coil 1				Coil 2				Coil 3			
	Deflection on D'Arsonval	E.M.F. in 10⁻⁶ volts per turn	Ampères per sq. cm.	Watts per sq. cm. in 10⁻⁶	Deflection on D'Arsonval	E.M.F. in 10⁻⁶ volts per turn	Ampères per sq. cm.	Watts per sq. cm. in 10⁻⁶	Deflection on D'Arsonval	E.M.F. in 10⁻⁶ volts per turn	Ampères per sq. cm.	Watts per sq. cm. in 10⁻⁶
6	3	12·9	·125	..	42	350	1·35	..	16	500	1·20	..
8	4	17·2	·166	2·86	30	250	·965	241	− 8	250	·602	150
10	0	0	0	0	0	0	0	0	+10	312	·752	235
11	..	..	..	..	..	..	..	..	..	..	..	..
12	+5·5	23·7	·229	5·43	+23	192	·74	142	+42	1310	3·16	4140
13	..	..	..	..	..	..	..	..	..	..	..	..
14	+4	17·2	·166	2·86	+24	200	·772	154	+28	·874	2·11	1840
15	..	..	..	..	..	..	..	..	..	..	..	..
16	+3·5	15·1	·145	2·19	+42	350	1·35	472	+19	593	1·43	848
17	..	..	..	..	..	..	..	..	..	..	..	..
18	+4	17·2	·166	2·86	+35	292	1·13	330	+ 9	281	·677	190
19	..	..	..	..	..	..	..	..	..	..	..	..
20	+2	8·6	·083	·714	+ 8	66·7	·257	17·1	0	0	0	0
20·5	..	..	..	..	..	..	..	..	..	..	..	..
22	− 5	21·5	·208	4·47	− 24	200	·772	154	42	1310	3·16	4140
24	− 4	17·2	·166	..	25	209	·804	..	30	937	2·26	..
26	− 3	..	·125	..	41	..	1·32	..	22	..	1·65	..

TABLE VII.—continued.

Time in seconds	Average induction per sq. cm. of Coil 1			Current in copper coils			Resultant force, taken from the given cyclic curve. H	Magnetising force due to induced currents			Energy dissipated by induced currents
	Area of E.M.F. curve in $\frac{1}{25}$ diagram squares	Half max. area ± other areas in sq. cm.	Induction per sq. cm. in c.g.s. units. B	Deflection on D'Arsonval	Ampères	Force per cm. linear. H_1		H_2	Areas required from curves in Fig. 18A, in diagram squares	Areas obtained from curves in Fig. 18A, as drawn, in diagram squares	Watts per cm. in 10^{-6}
6	..	..	..	− 39	·235	− 4·86					
8	..	..	..	− 32	·193	− 3·99					
10	0·0	+ 8·25	+ 1630	− 10	·060	− 1·25	+ 1·32	·07	·22	·4	30
11	0·8	+ 7·45	+ 1470	..	..	..	+ ·75				
12	2·8	+ 5·45	+ 1075	+ 28	·169	+ 3·49	+ ·20	3·69	11·7	12·5	1420
13	5·0	+ 3·25	+ 641	..	..	..	− ·30				
14	6·9	+ 1·35	+ 266	+ 37	·223	+ 4·62	− ·45	4·17	13·3	13·0	1500
15	8·5	− ·25	+ 49	..	..	..	− ·60				
16	10·0	− 1·75	− 345	+ 39	·235	+ 4·86	− ·90	3·96	12·6	14·9	1680
17	11·7	− 3·45	− 680	..	..	..	− ·95				
18	13·3	− 5·05	− 996	+ 33	·199	+ 4·12	− 1·10	3·02	9·6	11·0	820
19	15·0	− 6·75	− 1330	..	..	..	− 1·22				
20	16·3	− 8·05	− 1590	+ 10	·060	+ 1·25	− 1·30	1·95	6·2	4·5	75
20·5	16·5	− 8·25	− 1630	..	..	..	− 1·32				
22	..	..	..	− 26	·157	− 3·24					
24	..	..	..	− 36	·217	− 4·49					
26	..	..	..	− 39	·235	− 4·86					

TABLE VIII.—Periodic Time, 80 seconds. Maximum Current in Copper Coils, 1·06 ampères.

Time in seconds	Coil 1				Coil 2				Coil 3			
	Deflection on D'Arsonval	E.M.F. in 10^{-6} volts per turn	Ampères per sq. cm.	Watts per sq. cm. in 10^{-6}	Deflection on D'Arsonval	E.M.F. in 10^{-6} volts per turn	Ampères per sq. cm.	Watts per sq. cm. in 10^{-6}	Deflection on D'Arsonval	E.M.F. in 10^{-6} volts per turn	Ampères per sq. cm.	Watts per sq. cm. in 10^{-6}
12	1·0	…	…	…	4	…	…	…	3	…	…	…
14	·5	…	…	…	1	…	…	…	1·5	…	…	…
16	0	0	0	…	0	0	0	…	·5	…	…	…
18	0	0	0	…	0	0	0	…	0	…	…	…
20	0	0	0	0	·5	8·78	·0338	0	0	0	0	0
22	0	0	0	…	·5	8·78	·034	…	·5	22·1	·0532	…
24	0	0	0	…	1·0	17·6	·068	…	1	44·1	·106	…
26	·5	4·12	·0398	·164	1·0	17·6	·068	…	2	88·3	·213	…
28	·5	4·12	·0398	·164	2	35·1	·135	…	2	88·3	·213	…
30	1·0	8·25	·0795	…	3	52·7	·203	10·7	4	177	·426	…
32	1·5	12·4	·119	…	4	70·2	·271	…	5	221	·532	118
34	2·0	16·5	·159	2·62	8	140	·542	…	8	353	·851	…
36	4·0	33·0	·318	10·5	12	211	·812	171	14	618	1·49	…
38	5·0	41·2	·398	16·4	15	263	1·02	268	20	883	2·13	1880
40	4·0	33	·318	10·5	15	263	1·02	268	30	1320	3·19	4210
42	4·5	37·1	·358	13·3	45	790	3·05	2410	40	1770	4·26	7540
44	10	82·5	·795	65·6	51	896	3·45	3090	38	1680	4·04	6790
46	47	388	3·74	1451	45	790	3·05	2410	33	1460	3·51	5120
48	5·0	41·2	·308	16·4	35	615	2·37	…	28	1240	2·98	3690
50	·5	4·12	·0398	·16	10	176	·677	119	20	883	2·13	1880
52	·5	4·12	·0398	·16	2	35·1	·135	4·74	14	618	1·49	921
54	0	0	0	0	1	17·6	·068	1·20	4	177	·426	75
56	0	0	0	…	0	0	0	0	2	88·3	·213	18·8
58	0	0	0	…	0	0	0	…	·5	22·1	·053	·12
60	…	…	…	…	0	…	…	…	0	0	0	0
62	…	…	…	…	…	…	…	…	0	0	0	…
64	…	…	…	…	…	…	…	…	0	…	…	…
66	…	…	…	…	…	…	…	…	…	…	…	…
68	…	…	…	…	…	…	…	…	…	…	…	…
70	…	…	…	…	1	17·6	·068	…	1·5	…	…	…

TABLE VIII.—*continued.*

Time in seconds	Average induction per sq. cm. of Coil 1			Current in copper coils			Magnetising force due to induced currents		Resultant magnetising force per cm. linear. H	Energy dissipated by induced currents
	Area of E.M.F. curve in $\frac{1}{25}$ diagram squares	Half max. area ± other areas in sq. cm.	Induction per sq. cm. in C.G.S. units, B	Deflection on D'Arsonval	Ampères	Force per cm. linear. H_1	Area in diagram squares of curves in Fig. 19A from rad. 1·27 to 5·08	Force per cm. linear. H_2		Watts per cm. in 10^{-6}
12	...	...	...	49	1·04	+21·5				
14	...	...	...	45	0·95	+19·8				
16	0	...	...	39	·83	+17·1				
18	0	...	...	35	·74	+15·4	0	0	− 21·5	0
20	0	...	...	30	·64	+13·2				
22	·1	...	...	25	·53	+11·0				
24	·2	7·6	15,000	20	·424	+ 8·78				
26	·55	7·5	15,000	12	·254	+ 5·27				
28	·9	7·5	14,800	+ 2	·042	+ ·88				
30	1·3	7·4	14,600	− 6	·127	− ·88				
32	2·1	7·05	13,900	14	·297	− 2·63				
34	2·8	6·7	13,200	20	·424	− 6·15				
36	3·5	6·3	12,400	26	·55	− 8·78	5	+ 1·6	− 10·4	125
38	4·6	5·5	10,800	30	·64	− 11·4				
40	9·65	4·8	9,470	35	·74	− 13·2				
42	14·8	4·1	8,090	39	·83	− 15·4	13·8	+ 4·34	− 1·71	1,600
44	15·1	+3·0	5,920	44	·93	− 17·1	23·5	+ 7·39	+ 1·39	3,200
46	15·18	−2·05	4,040	48	1·02	− 19·3	...	...	...	5,070
48	15·2	7·2	14,200	49	1·04	− 21·1	42·0	+13·2	+ 2·2	10,400
50	15·2	7·5	14,800	50	1·06	− 21·5	45·0	+14·2	+ 2·9	10,500
52		7·58	14,950	48	1·02	− 21·9	10·0	+ 3·1	+ 17·2	8,750
54		7·6	15,000	43	·91	− 21·1	3·5	+ 1·1	+ 20·0	7,880
56		7·6	15,000	40		− 18·9	1·6	+ ·5	+ 21	625
58				35			·15	+ ·05	+ 21·8	
60							0	+ 0	...	0
62										
64										
66										
68										
70										

TABLE IX.—*Periodic Time, 20 seconds. Maximum Current in Copper Coils, 1·06 ampères*

Time in seconds	Coil 1				Coil 2				Coil 3			
	Deflection on D'Arsonval	E.M.F. in 10⁻⁶ volts per turn	Ampères per sq. cm.	Watts per sq. cm. in 10⁻⁶	Deflection on D'Arsonval	E.M.F. in 10⁻⁶ volts per turn	Ampères per sq. cm.	Watts per sq. cm. in 10⁻⁶	Deflection on D'Arsonval	E.M.F. in 10⁻⁶ volts per turn	Ampères per sq. cm.	Watts per sq. cm. in 10⁻⁶
26	10	109	1·05	··	23	1630	6·28	··	14	1920	4·63	··
28	50	545	5·25	··	+10	708	2·73	··	+ 3	412	·992	··
30	0	0	0	0	− 2	142	·546	77·5	− 8	1098	2·64	2,900
31												
32	5	54·5	·525	28·6	7	495	1·91	945	30	4120	9·92	40,900
33	4	43·6	·420	18·3	25	1770	6·82	12,100	24	3290	7·93	26,100
34												
35	10	109	1·05	114	25	1770	6·82	12,100	15	2060	4·96	10,200
36												
37	50	545	5·25	2860	10	708	2·73	1930	− 5	686	1·65	1,130
38												
39	0	0	0	0	0	0	0	0	+ 5	686	1·65	1,130
40												
42	5	54·5	·525	··	5	354	1·36	··	30	4120	9·92	··
44	4	43·6	·420	··	7	495	1·91	··	22	3020	7·27	··

TABLE IX.—*continued.*

Time in seconds	Average induction per sq. cm. of Coil 1			Current in copper coils			Magnetising force due to induced currents		Resultant magnetising force per cm. linear. H	Energy dissipated by induced currents. Watts per cm. in 10^{-6}
	Area of E.M.F. curve in ⅕ diagram squares	Half max. area ± other areas in sq. cm.	Induction per sq. cm. in c.g.s. units. B	Deflection on D'Arsonval	Ampères	Force per cm. linear. H_1	Area in diagram squares of curves in Fig. 20A from rad. 1·27 to 5·08	Force per cm. linear. H_2		
26				39	·86	17·8				
28				30	·430	+ 8·31				
30	0	+ 7·85	+ 15,500	+ 15			6·5	4·09	13·0	1,250
31				− 5	·143	− 2·97				
32	·5	+ 7·35	+ 14,500	25	·716	14·8	·22	13·8	10·0	20,000
33										
34	1·5	+ 7·35	+ 12,500	37	1·06	22·0	38	23·9	9·1	44,000
35										
36	2·8	+ 5·05	+ 9,950	30	·86	17·8	41	25·8	+ 3·8	33,700
37										
38	9·4	− 1·55	− 3,060		·43	8·91	+ 25	15·7	− 2·1	12,000
39				− 15						
40	15·7	− 7·85	− 15,500	+ 10	·287	5·94	− 1·5	·74	9·8	175
42				30	·86	17·8				
44										

38.

ON THE RUPTURE OF IRON WIRE BY A BLOW.

[From the *Proceedings of the Manchester Literary and Philosophical Society*, Vol. XI. pp. 40—45, 1872.]

THE usual method of considering the effect of impulsive forces, though in most cases very convenient, sometimes hides what a more ultimate analysis reveals. The following is an attempt to investigate the effect the blow of a moving mass has on a solid body in one or two simple cases; I venture to lay it before the Society on account of its connexion with the question of the strength of iron at different temperatures.

I assume the ordinary laws concerning the strains and stresses in an elastic solid to be approximately true, and that if the stress at any point exceed a certain limit rupture will result. Take the case of an elastic wire or rod, natural length l, modulus E, fixed at one end: the other end is supposed to become suddenly attached to a mass M moving with velocity V, which the tension of the wire brings to rest. The wire is thus submitted to an impulsive tension due to the momentum MV, and according to the usual way of looking at the subject of impact, the liability to rupture should be independent of l and proportional to MV. But in reality the mass MV is pulled up gradually, not instantaneously, and the wire is not at once uniformly stretched throughout, but a wave of extension or of tension is transmitted along the wire with velocity a, where $a^2 = \dfrac{E}{\mu}$ (μ being the mass of a unit of length of the wire); in an infinite wire this wave would be most intense in front. In the wire of length l this wave is reflected at the fixed point, and returns to the point of attachment of the mass M, and the effects of the direct and reflected waves must be added, and again we must add the wave as reflected from M back towards the fixed point. The question then of the breaking of the wire is very complicated, and may depend not merely on the strength of the material to resist rupture, but also on a, E, and l, and on M and V independently, not only on the product MV.

First take the case of an infinite wire; let x be the unstretched distance of any point from the initial position of the extremity which is fast to M, $x + \xi$ the distance of the same point from this origin at time t. The equation of motion is

$$(1) \quad \frac{d^2\xi}{dt^2} = a^2 \frac{d^2\xi}{dx^2},$$

and we have the condition

$$(2) \quad M \frac{d^2\xi}{dt^2} = E \frac{d\xi}{dx} \quad \text{when} \quad x = 0.$$

The general solution of (1) is $\xi = f(at - x)$.

Substitute in (2) and put $x = 0$.

$$Ma^2 f''(at) = - Ef'(at); \quad \text{but} \quad a^2 = \frac{E}{\mu}.$$

Therefore $\qquad Mf'(at) = -\mu f(at) - \dfrac{MV}{a};$

for initially

$$f(at) = 0 \quad \text{and} \quad f'(at) = -\frac{V}{a}.$$

Therefore $\qquad \dfrac{Maf'(at)}{\mu f(at) + \dfrac{MV}{a}} = -a,$

$$\mu f(at) + \frac{MV}{a} = \frac{MV}{a} \epsilon^{-\frac{\mu a t}{M}}.$$

Therefore $\qquad \xi = -\dfrac{MV}{\mu a}\left(1 - \epsilon^{-\frac{\mu (at - x)}{M}}\right)$

true at any point after $t > \dfrac{x}{a}$.

$$\text{Tension} = E \frac{d\xi}{dx} = \frac{VE}{a} \epsilon^{-\frac{\mu}{M}(at - x)}.$$

This is greatest when $at - x = 0$, and then $= V\sqrt{E\mu}$.

So that for the case of an infinite wire it will break unless the statical breaking force $> V\sqrt{E\mu}$; a limit wholly independent of M.

This result is approximately true in the case of a very long wire : if F be the force which acting statically would break the wire,

$$\text{velocity necessary} = \frac{F}{\sqrt{E\mu}}.$$

Any change then which increases E will render such a wire more liable to break under impact : cold has this effect ; we arrive then at the apparently anomalous result that though cold increases the tensile strength of iron, yet owing to increasing its elasticity in a higher ratio it renders it more liable to break under impact.

Now let us return to the case of the wire length l. We have the additional condition that when $x = l$, $\xi = 0$ for all values of t, and this will introduce a number of discontinuities into the solution. Up to the time $\dfrac{2l}{a}$ we may deduce the solution from the previous case ; from $t = 0$ to $t = \dfrac{l}{a}$ we have as before

$$(3) \quad \xi = \frac{MV}{\mu a}\left(\epsilon^{-\frac{\mu\,(at-x)}{M}} - 1\right);$$

but then reflection occurs, and we have

$$(4) \quad \xi = \frac{MV}{\mu a}\left\{\epsilon^{-\frac{\mu(at-x)}{M}} - \epsilon^{-\frac{\mu(at-2l+x)}{M}}\right\}.$$

It is to be observed that at any point x equation (3) applies from $t = \dfrac{x}{a}$ till $t = \dfrac{2l-x}{a}$, whilst (4) applies from $t = \dfrac{2l-x}{a}$ to $t = \dfrac{2l+x}{a}$.

I will not go into the question of the reflection at the mass M, but notice that when the wave is reflected at the fixed point

$$\frac{d\xi}{dx} = 2\,\frac{V}{a}.$$

Therefore tension $= 2V\sqrt{E\mu}$ or double our previous result.

We infer, then, that half the velocity of impact needed to break the wire near the mass is sufficient to break it at the fixed point, but that in both cases the breaking does not depend on the mass.

These results were submitted to a rough experiment. An iron wire, No. 13 gauge, about 27 feet long, and capable of carrying $3\frac{1}{2}$ cwt. dead weight, was seized in a clamp at top and bottom ; the top clamp rested on beams on an upper floor, whilst the lower served to receive the impact of a falling mass. The wire was kept taut by a 56 lb. weight hung below the lower clamp. The falling weight was a ball having a hole drilled in it sliding on the wire. It is clear that, although the clamp held without slipping, the blow must pass through it, and will be deadened thereby, so giving an advantage to the heavy weight. If the wire breaks some way up the wire, or at the upper clamp, it may be considered that the wire near the lower clamp stood the first onset of the blow, and hence that if the wire had been long enough it would have stood altogether.

I first tried $7\frac{1}{4}$ lbs.; the wire stood the blow due to falls of 6' and 6' 6'' completely, but broke at the lower clamp with 7' 0'' and 7' 2''. We may take 6' 9'' as the breaking height. With a 16 lb. weight dropped 5' 6'' the wire broke at the upper clamp. A 28 lb. was then tried; falls of 2' and 3' respectively broke it near the upper clamp; 4' 6'' broke it three feet up the wire in a wounded place; 5' broke it at the top clamp, and 6' was required to break it at the lower clamp. This may be taken as a rough confirmation of the result that double the velocity is required to break it at the lower clamp to that required to cause rupture at the upper. Lastly, 41 lbs. was tried; a fall of 4' 6'' broke it at the upper clamp, 5' 6'' at the lower; take 5' as height required to break at the lower.

In problems of this kind it has been assumed by some that two blows were equivalent when their vis vivas were equal, by others when the momenta were equal; my result is that they are equal when the velocities or heights of fall are equal.

Taking the 41 lbs. dropped 5' as a standard, since it will be least affected by the clamp, I have taken out the heights required for the other weights. Column 1 is the weight in lbs.; 2, the fall observed; 3, the fall required on vis viva theory; 4, that required by momentum theory:

(1)	(2)	(3)	(4)
41	5' 0''	5' 0''	5' 0''
28	6' 0''	7' 4''	10' 9''
16	5' 6''	12' 11''	33' 0''
$7\frac{1}{4}$	6' 9''	28' 3''	160'

It will be seen that the law here arrived at is the nearest of the three, besides which its deviation is accounted for by the deadening effect of the clamp.

But it remains to be explained why the $7\frac{1}{4}$ lbs. weight could not break the wire at the top at all, whereas the 28 lbs. broke it with a fall of only 2 feet. We should find some means of comparing the searching effect of two blows. For this we must look to friction.

Assuming that the friction between two sections of the wire is proportional to their relative velocity, a hypothesis which accounts well for certain phenomena in sound, I worked out its effect in this case, but the result failed to account for the facts. This should not be surprising, for though this assumption may be true or nearly so for small relative velocities, it may well fail here when they are large. The discrepancy may perhaps be attributed to the fact that a strain which a wire will stand a short time, will ultimately break it, and possibly in part to want of rigidity in the supports of the upper clamp, both of which would favour the heavy weight.

I think we may conclude from the above considerations and rough experiments—

1st. That if any physical cause increase the tenacity of wire, but increase the product of its elasticity and linear density in a more than duplicate ratio, it will render it more liable to break under a blow.

2nd. That the breaking of wire under a blow depends intimately on the length of the wire, its support, and the method of applying the blow.

3rd. That in cases such as surges on chains, etc., the effect depends more on the velocity than on the momentum or vis viva of the surge.

4th. That it is very rash to generalize from observations on the breaking of structures by a blow in one case to others even nearly allied, without carefully considering all the details.

39.

FURTHER EXPERIMENTS ON THE RUPTURE OF IRON WIRE.

[From the *Proceedings of the Manchester Literary and Philosophical Society*, Vol. XI. pp. 119—121, 1872.]

In a paper read before this Society some weeks ago I gave a theory of the rupture of an iron wire under a blow when the wire is very long, differing from that usually accepted practically, and an account of a few experiments in confirmation.

In the simple case considered mathematically, certain conditions which have a material effect on the result are wholly neglected, such as the weight hung below the clamp to keep the wire taut, and the mass and elasticity of the clamp; these I have taken into consideration.

Of course it is impossible to make experiments on an infinitely long wire; we are therefore compelled to infer the breaking blow for such a wire from the blow required to break a short wire *close* to the clamp. The wire used in the following experiments was from 9 to 12 feet long, the clamp weighed 26 oz., and the weight at the end of the wire was 61 lbs. Several attempts were made to support the upper extremity of the wire on an indiarubber spring, in order that the wire might behave like a long wire and break at the bottom, and not be affected by waves reflected from the upper clamp, but without success; so that I was obliged to fall back on the plan of discriminating the cases in which the wire broke at the lower clamp from those in which the wave produced by the blow passed over this point without rupture and broke the wire elsewhere.

The height observed is corrected by multiplication by the factor $\left(\dfrac{M}{M+M'}\right)^2$ where M is the mass of the falling weight and M' of the clamp. This correction rests on the assumption that the clamp and cast iron weight are practically incompressible, and hence that at the moment of impact they take a common velocity which is that causing rupture of the wire. This assumption will of course be slightly in error, and experiments were made in which leather washers were interposed between the clamp and the iron weight to cushion the blow. The error produced by these washers would be of the same nature as that produced by elasticity in the clamp, but obviously many times as large. If the error produced by one thick leather washer be but 10 inches of reduced height, surely the effect of the elasticity of the clamp will fall well within the limits of error in these experiments.

The effect of cold on the breaking of the wire was tried thus— the clamp and the lower extremity of the wire were cooled by means of ether spray, and the weight dropped as before. The effect of cooling the wire near the clamp was in all cases to make the wire break more easily, in some cases very markedly so. A similar result would follow under similar circumstances from the formula for the resilience $\frac{1}{2}\dfrac{lF^2}{E}$; and it is the almost universal experience of those who have to handle crane chains and lifting tackle that these are most liable to breakage in cold weather. To this effect of temperature and to the variable quality of wire even in the same coil I attribute the discrepancy between the various observations.

The first column gives the height of fall observed, the second the reduced height, and the third the point at which the wire broke. The observations marked * are those in which cold was applied. The two series were tried on different days about a fortnight apart and on wire from different parts of the same coil. In all cases the upper clamp rested on the bare boards of the floor above.

First Series.

16 lbs. weight.

Inches.	Inches.	Point of Rupture.
72	60	18″ from top.
78	65	12″ from bottom.
78	65	24″ from top.
81	67½	at top and bottom.
82	68½	21″ from top.
84	70	at bottom.
84	70	at bottom.
*48	40	did not break.
*54	45	at bottom.
*60	50	at bottom.
*72	60	at bottom.

28 lbs. weight.

72	65	20″ from top.
78	70	close to top.
79½	71½	at bottom.
81	73	at bottom.

7 lbs. weight.

81	54	at top.
84	56	at bottom.
*72	48	at bottom.
*75	50	at bottom.

Second Series.

28 lbs. weight.

54	48	broke at top.
60	53½	bottom and half-way up.
60	53½	at top.
63	56	at bottom.
66	59	at bottom.
69	61½	at bottom.
72	64½	at bottom.
*36	32	at top.
*48	43	at bottom.

16 lbs. weight.

Inches.	Inches.	Point of Rupture.
60 ………	50 ………	half-way up.
66 ………	55 ………	at bottom.

With one dry leather washer.

72 ………	60 ………	4″ from bottom.
66 ………	55 ………	near top.

Two dry washers.

72 ………	60 ………	6″ from bottom.

Three soaked washers.

78 ………	65 ………	broke in middle.
83 ………	69 ………	at top.

It should be noticed that the formula velocity $= \dfrac{F}{\sqrt{E\mu}}$ cannot be depended on except as indicating the general character of the phenomena; for let us attempt to deduce the height of fall from this formula, $h = \dfrac{1}{2g}\dfrac{F^2}{E\mu}$.

An inch wire 1 foot long weighs 3·34 lbs., the breaking force in proper units $= 80,000 \times 32$, and the elasticity $= 25,000,000 \times 32$, whence $h = 38$ feet about.

This discrepancy I have not yet accounted for.

40.

THE MATHEMATICAL THEORY OF TARTINI'S BEATS.

[From the *Messenger of Mathematics*, New Series, No. 14, 1872.]

WHEN two musical sounds of different pitch are produced together in sufficient intensity, a third and faint musical sound may be observed making a number of vibrations equal to the difference of the numbers of vibrations of the two notes sounded.

This phenomenon was first observed by Sorge, a German organist, in 1740, and shortly afterwards discovered independently by Tartini. Dr Young offered an explanation, according to which the note had no real objective existence, but was to be attributed to the organ of hearing itself. When two musical sounds differ in pitch slightly, their interference causes a continual rising and falling of the intensity of the sound, it produces what are known as beats; as the beats become more rapid they are more difficult to distinguish from each other, and Young supposed that when they attain a certain number the ear associates them together and makes a distinct musical note of them, just as it does from a series of ordinary sound waves when they become sufficiently numerous to affect the ear. On this theory the sound does not exist as a wave in the air at all, but first arises in the conscious- ness of the observer, and it should be impossible to intensify the effect by outward appliances, such as resonators, except by inten- sifying the sounds which give rise to it. Helmholtz has shown experimentally that these tones may be intensified by resonators and even detected by membranes capable of vibrating in unison with the "Resultant Sound," without the aid of the ear at all. And he has shown that their existence, as well as those of other "Resultant Sounds," may be explained by considering the squares and higher powers of the amplitude of vibration in the equations expressing the motion of the substance conveying the waves of sound.

The following does not pretend to anything new in principle, but is simply an application of a method, essentially the same as that used by Mr Earnshaw in a paper read before the Royal Society in 1860, to solve the equations of motion to the second order, and so work out the explanation given by Helmholtz.

Let us consider the motion of the air when a series of plane waves are passing through it. Let the axis of x be perpendicular to the plane of the waves, and let the velocity of particles, whose undisturbed position is defined by $x = 0$, be

$$v = \frac{d\xi}{dt} = \Sigma A \sin (mt + \alpha),$$

i.e. be the sum of any number of harmonic terms. We propose to find the motion of the particles of air at distance x from the plane $x = 0$. Let ϕ be the characteristic function of the motion, then $\frac{d\phi}{dx}$ is the velocity of the particles in the plane distant x from the origin.

Let p be pressure and ρ density at any point at any time, γ the ratio of specific heats under constant pressure and volume, and σ the mean value of ρ.

Then the equation of motion may be written

$$\frac{d^2\phi}{dt^2} + 2 \frac{d\phi}{dx} \cdot \frac{d^2\phi}{dx\,dt} + \left(\frac{d\phi}{dx}\right)^2 \frac{d^2\phi}{dx^2}$$

$$= a^2 \frac{d^2\phi}{dx^2} - (\gamma - 1) \left(\frac{d\phi}{dt} + \frac{1}{2} \overline{\frac{d\phi}{dx}}^2\right) \frac{d^2\phi}{dx^2} \quad \ldots\ldots\ldots(1),$$

where $\qquad\qquad a^2 = \gamma\kappa\sigma^{\gamma-1},$

as may be shown by combining the equation $p = \kappa\rho^\gamma$ with the usual equation of motion and continuity.

The following will be found on trial to be solutions of (1):

$$\frac{d\phi}{dx} = F\left[x - \left\{ a + \frac{\gamma + 1}{2} \frac{d\phi}{dx} \right\} t \right] \quad\ldots\ldots\ldots\ldots(2),$$

$$\frac{d\phi}{dx} = f\left[t - \frac{x}{a + \dfrac{\gamma + 1}{2} \dfrac{d\phi}{dx}} \right] \quad\ldots\ldots\ldots\ldots(3).$$

The former is suitable when the velocity at all points is known for some given time, and the latter, when it is given at all times for some fixed point. The physical meaning of the equations is the same. The latter is suitable for our purpose; let us take it for verification.

Assume

$$\frac{d\phi}{dx} = f\left\{ t - \frac{x}{a + \mu \dfrac{d\phi}{dx}} \right\}$$

with the view of determining μ.

Eliminating f, we have

$$\frac{d^2\phi}{dx\,dt} + \left(a + \mu \frac{d\phi}{dx} \right) \frac{d^2\phi}{dx^2} = 0 \quad\ldots\ldots\ldots\ldots(4).$$

Integrating with respect to x

$$\frac{d\phi}{dt} + a \frac{d\phi}{dx} + \frac{\mu}{2} \left(\frac{d\phi}{dx} \right)^2 = 0 \quad\ldots\ldots\ldots\ldots(5).$$

Differentiating with respect to t

$$\frac{d^2\phi}{dt^2} + a \frac{d^2\phi}{dx\,dt} + \mu \frac{d\phi}{dx} \frac{d^2\phi}{dx\,dt} = 0 \quad\ldots\ldots\ldots\ldots(6).$$

To (6) add (4) multiplied by $-\left\{ a + (\mu - 2) \dfrac{d\phi}{dx} \right\}$, and (5) multiplied by $2\,(\mu - 1) \dfrac{d^2\phi}{dx^2}$; the resulting equation is identical with (1), provided $\mu = \dfrac{\gamma + 1}{2}$.

Now in (3) putting $x = 0$, we have

$$f(t) = \Sigma A \sin (mt + \alpha).$$

Therefore the complete solution of the problem proposed is contained in the equation

$$\frac{d\phi}{dx} = \Sigma A \sin\left\{ m\left(t - \frac{x}{a + \dfrac{\gamma+1}{2}\dfrac{d\phi}{dx}} \right) + \alpha \right\} \quad \ldots\ldots\ldots(7):$$

from this $\dfrac{d\phi}{dx}$ must be expressed in a series of harmonic terms of t, in order that we may determine the musical character of the vibrations at any point, at distance x from the origin. It is clear that they will be, as is well known, of the form $\dfrac{\sin}{\cos}(pm \pm qn)\,t$, and that the coefficient cannot be of a lower than the $(p+q)^{\text{th}}$ order; it remains, by expanding (7), to find the coefficients of the various terms. We will do so for the second order alone

$$\frac{d\phi}{dx} = \Sigma \left[A \sin\left\{ m\left(t - \frac{x}{a}\right) + \alpha \right\} \right.$$

$$+ A\frac{d\phi}{dx} m \frac{\gamma+1}{2} \frac{x}{a^2} \cos\left\{ m\left(t - \frac{x}{a}\right) + \alpha \right\}$$

$$\left. + \text{terms of third and higher orders which we neglect} \right],$$

$$\frac{d\phi}{dx} = \Sigma A \sin\left\{ m\left(t - \frac{x}{a}\right) + \alpha \right\}$$

$$+ \Sigma A m \frac{\gamma+1}{2} \frac{x}{a^2} \cos\left\{ m\left(t - \frac{x}{a}\right) + \alpha \right\} . \Sigma A \sin\left\{ m\left(t - \frac{x}{a}\right) + \alpha \right\} :$$

thus we see that from terms of the second order we shall have the following notes arise in the propagation of the sound which are not present at the origin of the disturbance.

1st.   From any term $A \sin mt$ will arise

$$m A^2 \frac{\gamma+1}{4} \frac{x}{a^2} \sin 2mt.$$

2nd.   From any pair of terms $A \sin mt$ and $B \sin nt$ will arise two resultant tones, one higher than its components, the other lower, viz.:

$$A B \frac{\gamma+1}{4} \frac{x}{a^2} \{(m+n)\sin(m+n)\,t - (m-n)\sin(m-n)\,t\},$$

which gives the numerical value of Tartini's Beat at any point.

41.

ON THE STRESSES PRODUCED IN AN ELASTIC DISC BY RAPID ROTATION.

[From the *Messenger of Mathematics*, New Series, No. 16, 1872.]

LET a plane circular lamina rotate about an axis through its centre perpendicular to its plane, it is proposed to find the stress produced thereby at all points of the lamina. This problem has a certain practical value. It not unfrequently happens that the grindstones used for polishing metal work are ruptured by the tensions caused by rapid rotation, portions of the stone being projected with such velocity as to cause serious injury and even loss of life. It is of importance, therefore, to determine the comparative velocities which stones of various sizes will stand, and the lines along which they are most liable to fracture.

Let a be the radius of the plate, and b of a hole cut in the centre, σ the mass of a unit of area, ω the angular velocity of rotation, r be the distance of any point P of the lamina when the disc is at rest, $r + \rho$ when the disc is in rotation. The strains about P will be $\dfrac{d\rho}{dr}$ along the radius, and $\dfrac{\rho}{r}$ perpendicular to the radius. And the consequent stresses respectively $A\,\dfrac{d\rho}{dr} + B\,\dfrac{\rho}{r}$ and $B\,\dfrac{d\rho}{dr} + A\,\dfrac{\rho}{r}$, where A and B are constants dependent on the nature of the material.

Consider the motion of an element enclosed by two consecutive radii and two circles r and $r + dr$; resolving along one of the bounding radii we have

$$\frac{d}{dr} r \left(A \frac{d\rho}{dr} + B \frac{\rho}{r} \right) - \left(B \frac{d\rho}{dr} + A \frac{\rho}{r} \right) + \sigma r \omega^2 = 0,$$

or

$$A \left(\frac{d^2\rho}{dr^2} + \frac{1}{r} \frac{d\rho}{dr} - \frac{\rho}{r^2} \right) + \sigma \omega^2 = 0 \quad \dots\dots\dots\dots(1).$$

The complete solution of this equation is

$$\rho = C_1 r + \frac{C_2}{r} - \frac{\sigma \omega^2 r^2}{3A},$$

whence radial stress at r

$$= C_1 (A + B) - \frac{C_2 (A - B)}{r^2} - \frac{2A + B}{3A} \sigma \omega^2 r.$$

The constants C_1 and C_2 must be determined so that this expression may vanish alike when $r = b$ and when $r = a$. This is the case if

$$C_1 = \frac{(2A + B) \sigma \omega^2}{3A (A + B)} \cdot \frac{a^3 - b^3}{a^2 - b^2},$$

$$C_2 = \frac{(2A + B) \sigma \omega^2}{3A (A - B)} \cdot \frac{a^2 b^2 (a - b)}{a^2 - b^2}.$$

Hence, radial stress

$$= \frac{(2A + B) \sigma \omega^2}{3A} \left\{ \frac{a^3 - b^3}{a^2 - b^2} - \frac{a^2 b^2 (a - b)}{(a^2 - b^2) r^2} - r \right\} \quad \dots\dots\dots(2),$$

and tangential stress

$$= \frac{(2A + B) \sigma \omega^2}{3A} \left\{ \frac{a^3 - b^3}{a^2 - b^2} + \frac{a^2 b^2 (a - b)}{(a^2 - b^2) r^2} - \frac{2B + A}{2A + B} r \right\} \quad \dots(3).$$

The radial stress is greatest when

$$r = \sqrt[3]{\frac{2a^2 b^2}{a + b}},$$

but since, for all substances $B < A$, the tangential stress at any point is the greater, and has its highest value

$$\left(\frac{2a^2}{a + b} + \frac{A - B}{2A + B} b \right) \frac{2A + B}{3A} \sigma \omega^2 \quad \dots\dots\dots\dots(4),$$

when $r = b$.

This highest value decreases as b increases.

The above solves the problem proposed. Let us see what we may conclude concerning the splitting, or, as it is commonly called, the " bursting " of grindstones.

1st. The stone will break with a radial fracture beginning at the inside. The expression " bursting " is then appropriate.

2nd. The greater the hole in the centre of the stone the stronger will the stone be. A solid stone runs at considerable disadvantage.

3rd. The proportion of the radius of stone to radius of hole being the same, the admissible angular velocity of stone varies inversely as the square root of the radius, and hence velocity of surface varies directly as the square root of the radius.

42.

ON THE EFFECT OF INTERNAL FRICTION ON RESONANCE.

[From the *Philosophical Magazine* for March 1873.]

As a typical case which may be taken as illustrating the nature of the phenomena in more complex cases, let us consider the motion of a string, of a column of air, or an elastic rod vibrating longitudinally, one extremity being fixed, whilst the other is acted on so that its motion is expressed by a simple harmonic function of the time.

Let l be the length of the string, a the velocity with which a wave is transmitted along it, ξ the displacement of a point of the string distant x from the fixed extremity at the time t. In the hypothetical case, in which there is no friction, no resistance of a surrounding medium, and the displacements are indefinitely small, the equation of motion is

$$\frac{d^2\xi}{dt^2} = a^2 \frac{d^2\xi}{dx^2} \dots\dots\dots\dots\dots\dots\dots(1),$$

with the conditions that at the extremities $\xi = 0$ when $x = 0$, and $\xi = A \sin nt$ when $x = l$, also that at some epoch ξ shall be a specified function of x.

If we start with the string straight and at rest, we have the condition $\xi = 0$ for all values of x from zero to very near l when $t = 0$, and we readily find

$$\xi = \frac{A}{\sin \dfrac{nl}{a}} \sin nt . \sin \frac{nx}{a} + \Sigma C_p \sin \frac{p\pi x}{l} . \sin \frac{p\pi at}{l} \dots\dots(2),$$

where $C_p = (-1)^p \dfrac{2nal}{n^2 l^2 - p^2 \pi^2 a^2}$.

When $\dfrac{nl}{a}$ is very nearly a multiple of π (*i.e.* when the note sounded by the forcing vibration at the extremity is almost the same as one of the natural notes of the string), we have two notes sounded with intensity, viz. one the same as the forcing vibration, the other native to the string. That this is the case may be readily seen with a two-stringed monochord, the strings being nearly in unison : one string being sounded, the motion of the other is seen by the eye to be intermittent, the period of variation being the same as that of the beats of the two strings sounded together. But should $\dfrac{nl}{a}$ be an exact multiple of π, two terms in the value of ξ become infinite, and our whole method of solution is invalid. A somewhat similar difficulty, of course, occurs in the lunar and planetary theories, but with this difference : there the difficulty is introduced by the method of solving the differential equation, and is avoided by modifying the first approximation to a solution ; here it is inherent in the differential equation, and can only be avoided by making that equation express more completely the physical circumstances of the motion. One or more of the assumptions on which the differential equation rests is invalid. We must look either to terms of higher orders of smallness, to resistance of the air, or to internal friction. With the modifications due to the last cause we are now concerned.

The approximate effect of internal friction is probably to add to the stress $E\dfrac{d\xi}{dx}$, produced by the strain $\dfrac{d\xi}{dx}$ when the parts of the body are relatively at rest, a term proportional to the rate at which the strain is changing ; so that the stress when there is relative motion will be $E\left(\dfrac{d\xi}{dx}+k\dfrac{d^2\xi}{dx\,dt}\right)$, and our equation of motion becomes

$$\frac{d^2\xi}{dt^2}=a^2\left(\frac{d^2\xi}{dx^2}+k\frac{d^3\xi}{dx^2\,dt}\right) \quad\ldots\ldots\ldots\ldots\ldots(3).$$

The solution of this equation will contain two classes of terms. First, a series corresponding to those under the sign of summation in (2), which principally differ from (2) in the coefficients decreasing in geometrical progression with the time, the highest fastest, and in the total absence of the notes above a certain order as periodic terms ; these terms we may consider as wholly resulting from the initial conditions, and as having no permanent effect on the motion. Second, a term corresponding to the first term of (2), and which expresses the state of steady vibration when work enough is continually done by the forced vibration of the extremity to maintain a constant amplitude. The investigation of this term is a little more troublesome, because the motion is periodic, the effect of friction being to alter the motion in a manner dependent on the position of the point, not on the time, and equation (3) cannot be satisfied by a sine or a cosine alone of the time.

Assume $\quad \xi = \phi(x) \sin mt + \psi(x) \cos mt,$

or a series of such terms, if possible, each pair satisfying equation (3). Substitute in the equations of motion, and equate coefficients of $\sin mt$ and $\cos mt$,

$$a^2 (\phi'' - km\psi'') = -m^2\phi \atop a^2 (\psi'' + km\phi'') = -m^2\psi \Big\} \quad \dots\dots\dots\dots(4).$$

Assume

$$\phi = c_1 \sin \lambda x \atop \psi = c_2 \sin \lambda x \Big\} \quad \dots\dots\dots\dots\dots(5),$$

where c_1, c_2, and λ may be imaginary, but ϕ and ψ are real: this form is indicated as suitable, because ξ must change sign with x.

We obtain

$$a^2 (c_1 - c_2 km) \lambda^2 = m^2 c_1 \atop a^2 (c_2 + c_1 km) \lambda^2 = m^2 c_2 \Big\} \quad \dots\dots\dots\dots(6),$$

whence

$$c_1^2 = -c_2^2, \quad c_2 = \pm c_1 \sqrt{-1} \dots\dots\dots\dots(7),$$

and

$$\lambda = \pm \mu \left(1 \pm \sqrt{-1} \tan \frac{\theta}{2}\right),$$

where $\tan \theta = km$,

$$\mu = \frac{m}{a} \frac{\cos \frac{\theta}{2}}{\sqrt[4]{1 + k^2 m^2}}.$$

The most general real expression for ϕ is then

$$\frac{A_1 + B_1 \sqrt{-1}}{2} \sin \mu \left(1 + \sqrt{-1} \tan \frac{\theta}{2}\right) x$$

$$+ \frac{A_1 - B_1 \sqrt{-1}}{2} \sin \mu \left(1 - \sqrt{-1} \tan \frac{\theta}{2}\right) x ;$$

or, as it may be written,

$$\phi = A_1 \sin \mu x \cdot \frac{\epsilon^{\mu \tan \frac{\theta}{2} \cdot x} + \epsilon^{-\mu \tan \frac{\theta}{2} \cdot x}}{2}$$
$$+ B_1 \cos \mu x \cdot \frac{\epsilon^{\mu \tan \frac{\theta}{2} \cdot x} - \epsilon^{-\mu \tan \frac{\theta}{2} \cdot x}}{2} \qquad \Bigg\} \quad \dots\dots(8).$$

Similarly

$$\psi = A_2 \sin \mu x \cdot \frac{\epsilon^{\mu \tan \frac{\theta}{2} \cdot x} + \epsilon^{-\mu \tan \frac{\theta}{2} \cdot x}}{2}$$
$$+ B_2 \cos \mu x \cdot \frac{\epsilon^{\mu \tan \frac{\theta}{2} \cdot x} - \epsilon^{-\mu \tan \frac{\theta}{2} \cdot x}}{2}.$$

The constants will be connected by the relations

$$\begin{cases} A_1 + B_1 \sqrt{-1} = A_2 \sqrt{-1} - B_2, \\ A_1 - B_1 \sqrt{-1} = - A_2 \sqrt{-1} - B_2; \end{cases}$$

that is,

$$A_1 = - B_2 \text{ and } B_1 = A_2 \qquad \dots\dots\dots\dots(9).$$

Let

$$P = \sin \mu l \cdot \frac{\epsilon^{\mu \tan \frac{\theta}{2} \cdot l} + \epsilon^{-\mu \tan \frac{\theta}{2} \cdot l}}{2}$$
$$Q = \cos \mu l \cdot \frac{\epsilon^{\mu \tan \frac{\theta}{2} \cdot l} - \epsilon^{-\mu \tan \frac{\theta}{2} \cdot l}}{2} \qquad \Bigg\} \quad \dots\dots(10).$$

If possible, let m be other than n; when $x = l$, we have $\phi = 0$ and $\psi = 0$, or

$$\begin{cases} A_1 P + B_1 Q = 0, \\ B_1 P - A_1 Q = 0; \end{cases}$$

therefore, since A_1, B_1 must be real, they must vanish, and we conclude that the only steady vibration is of the same period as that impressed on the extremity.

Let $m = n$; when $x = l$, $\phi = A$ and $\psi = 0$; hence

$$\begin{matrix} PA_1 + QB_1 = A \\ PB_1 - QA_1 = 0 \end{matrix} \Bigg\} ;$$

$$A_1 = \frac{AP}{P^2 + Q^2}$$
$$B_1 = \frac{AQ}{P^2 + Q^2} \qquad \Bigg\} \dots\dots\dots\dots\dots(11).$$

This completely determines the steady vibration of the string.

Suppose a change to take place in the forcing vibration, it is easy to see that the result will be that momentarily all the notes natural to the string with both ends fixed will be sounded. This conclusion could readily be tested by graphically describing the motion of a point of a string moving in the manner supposed, the motion being produced by a tuning-fork actuated by an electromagnet. If this be verified, an attempt might be made to determine the value of k for various strings or wires by comparing the amplitude of vibration at the points of greatest and least vibration; and at the different points of least vibration true nodes will not occur. The curve having x for abscissa, and the maximum value of ξ at each point for ordinate, might possibly be portrayed by photographing a vibrating string. The calculations would be much facilitated by the fact that $\mu = \dfrac{n}{a}$ if small quantities of the second order are neglected. Suppose that $\mu l = 2\pi$, a case of strong resonance; then $P = 0$ and $Q = \pi k n$ very nearly; we have $A_1 = 0$ and $B_1 = \dfrac{A}{\pi k n}$, and the motion is expressed by the equation

$$\xi = \frac{A}{\pi k n} \left\{ \frac{k n^2 x}{a} \cos \frac{nx}{a} \sin nt + \sin \frac{nx}{a} \cos nt \right\}.$$

Let the amplitudes observed at the node and middle of ventral segments of the string be α, β; we have

$$\left. \begin{aligned} \alpha &= \frac{A\,nl}{2\pi a} \\[2mm] \beta &= \frac{A}{\pi k n} \end{aligned} \right\} \dots\dots\dots\dots\dots\dots\dots(12);$$

therefore

$$k = \frac{2\alpha}{\beta} \frac{a}{n^2 l} = \frac{\alpha}{\beta} \frac{1}{\pi n},$$

the result being expressed in seconds. It is worth noticing that the vibrations throughout the ventral segments in this case are nearly a quarter of a vibration behind the extremity in phase. If the theory of friction here applied be correct, many important facts could follow from a determination of the value of k in different substances—for example, the relative duration of the harmonics of a piano-wire.

Let us now calculate what is the work done by the force maintaining the vibration of the extremity. The force there exerted is

$$E\left(\frac{d\xi}{dx} + k\,\frac{d^2\xi}{dx\,dt}\right),$$

and the work done in time dt is

$$E\left(\frac{d\xi}{dx} + k\,\frac{d^2\xi}{dx\,dt}\right)\frac{d\xi}{dt}\,dt,$$

x being put equal to l after differentiation. We have then work done from time 0 to time t

$$= \int_0^t \left\{E\left(\frac{d\xi}{dx} + k\,\frac{d^2\xi}{dx\,dt}\right)\frac{d\xi}{dt}\right\}_{x=l} dt.$$

In estimating the work done in any considerable period, we may exclude the periodic terms as unimportant. Hence work done on extremity of string

$$= \frac{nEt}{2}\left\{\left(\frac{d\psi}{dx} + kn\,\frac{d\phi}{dx}\right)\phi - \left(\frac{d\phi}{dx} - kn\,\frac{d\psi}{dx}\right)\psi\right\}_{x=l}$$

$$= \frac{nEt}{2}\left\{\frac{d\psi}{dx} + kn\,\frac{d\phi}{dx}\right\}_{(x=l)} A.$$

An expression for this could of course be at once written down without approximation; but the case where k is small is most important; then we have

$$\begin{cases} P = \sin\dfrac{nl}{a}, \\[2ex] Q = \cos\dfrac{nl}{a}\cdot\dfrac{n^2 lk}{2a}, \end{cases}$$

$$\begin{cases} A_1 = \dfrac{A}{\sin\dfrac{nl}{a}}, \\[3ex] B_1 = \dfrac{A\cos\dfrac{nl}{a}}{\sin^2\dfrac{nl}{a}}\cdot\dfrac{n^2 lk}{2a}, \end{cases}$$

unless $\sin \dfrac{nl}{a}$ becomes very small.

$$
\begin{cases}
\phi = A \dfrac{\sin \dfrac{nx}{a}}{\sin \dfrac{nl}{a}}, \\[3em]
\psi = \dfrac{A}{\sin^2 \dfrac{nl}{a}} \cdot \dfrac{kn^2}{2a} \left\{ l \cos \dfrac{nl}{a} \cdot \sin \dfrac{nx}{a} \right. \\[3em]
\qquad\qquad\qquad \left. - x \cos \dfrac{nx}{a} \sin \dfrac{nl}{a} \right\}.
\end{cases}
$$

Work done on the string

$$
= \dfrac{n^3 Etk}{4a \sin^2 \dfrac{nl}{a}} A^2 \left\{ l \cos^2 \dfrac{nl}{a} \cdot \dfrac{n}{a} - \cos \dfrac{nl}{a} \sin \dfrac{nl}{a} \right.
$$

$$
\left. + l \dfrac{n}{a} \sin^2 \dfrac{nl}{a} + 2 \sin \dfrac{nl}{a} \cos \dfrac{nl}{a} \right\}
$$

$$
= \dfrac{n^3 Etk A^2}{4a \sin^2 \dfrac{nl}{a}} \left\{ \dfrac{nl}{a} + \sin \dfrac{nl}{a} \cos \dfrac{nl}{a} \right\}.
$$

If

$$
\sin \dfrac{nl}{a} = 0, \quad Q = \pm \dfrac{n^2 lk}{2a},
$$

$$
A_1 = 0, \text{ and } B_1 = \pm \dfrac{2a}{n^2 lk} A,
$$

$$
\begin{cases}
\phi = \pm \dfrac{A}{l} x \cos \dfrac{nx}{a}, \\[2em]
\psi = \pm \dfrac{2aA}{n^2 lk} \sin \dfrac{nx}{a}.
\end{cases}
$$

Work done $= \dfrac{A^2}{lk} Et.$

We infer that the energy imparted to the string varies as the square of the amplitude of vibration of the extremity, that it rapidly increases as the period approaches that of the string, that, if these periods differ materially, the work is directly proportional to the friction and increases rapidly with the number of vibrations —but that if the periods are identical, the work varies *inversely* as the friction, the diminishing of the friction being more than counterbalanced by the increased amplitude.

It is interesting to examine how this energy is distributed over the string. This is easily done by writing down the work done by one portion of the string from x to l, on the remainder from 0 to a, and then taking the differential; we readily find that work absorbed by portion dx of string

$$= \frac{n^2 kEt}{2}\left(\overline{\frac{d\phi}{dx}}\Big|^2 + \frac{d\psi}{dx}\Big|^2\right) dx.$$

Substituting, we obtain, when the string does not resonate,

$$\text{work} = \frac{n^4 kEt}{2a}\frac{\cos^2\dfrac{nx}{a}}{\sin^2\dfrac{nl}{a}} A^2\, dx\,;$$

when the string resonates,

$$= \frac{Et}{l^2 k}\cos^2\frac{nx}{a}\cdot A^2 dx.$$

In either case the absorption of energy, and therefore the heating-effect, is greatest at the nodes, and, omitting squares of k, vanishes at the middle of the ventral segments. Directly the contrary will result from the friction of the string against the air.

43.

ON THE OPTICAL PROPERTIES OF A TITANO-SILICIC GLASS.—By Professor STOKES and J. HOPKINSON.

[From the *Report of the British Association for the Advancement of Science* for 1875.]

AT the Meeting of the Association at Edinburgh in 1871, Professor Stokes gave a preliminary account of a long series of researches in which the late Mr Vernon Harcourt had been engaged on the optical properties of glasses of a great variety of composition, and in which, since 1862, Professor Stokes had co-operated with him*. One object of the research was to obtain, if possible, two glasses which should achromatize each other without leaving a secondary spectrum, or a glass which should form with two others a triple combination, an objective composed of which should be free from defects of irrationality, without requiring undue curvature in the individual lenses. Among phosphatic glasses, the series in which Mr Harcourt's experiments were for the most part carried on, the best solution of this problem was offered by glasses in which a portion of the phosphoric was replaced by titanic acid. It was found, in fact, that the substitution of titanic for phosphoric acid, while raising, it is true, the dispersive power, at the same time produces a separation of the colours at the blue as compared with that at the red end of the spectrum, which ordinarily belongs only to glasses of a much higher dispersive power. A telescope made of disks of glass prepared by Mr Harcourt was, after his death, constructed for

---

* *Report* for 1871, Transactions of the Sections, p. 38.

Mrs Harcourt by Mr Howard Grubb, and was exhibited to the Mathematical Section at the late Meeting in Belfast. This telescope, which is briefly described in the 'Report*,' was found fully to answer the expectations that had been formed of it as to destruction of secondary dispersion.

Several considerations seemed to make it probable that the substitution of titanic acid for a portion of the silica in an ordinary crown glass would have an effect similar to what had been observed in the phosphatic series of glasses. Phosphatic glasses are too soft for convenient employment in optical instruments; but should titano-silicic glasses prove to be to silicic what titano-phosphatic glasses had been found to be to phosphatic, it would be possible, without encountering any extravagant curvatures, to construct perfectly achromatic combinations out of glasses having the hardness and permanence of silicic glasses; in fact the chief obstacle at present existing to the perfection of the achromatic telescope would be removed, though naturally not without some increase to the cost of the instrument. But it would be beyond the resources of the laboratory to work with silicic glasses on such a scale as to obtain them free from striæ, or even sufficiently free to permit of a trustworthy determination of such a delicate matter as the irrationality of dispersion.

When the subject was brought to the notice of Mr Hopkinson he warmly entered into the investigation; and, thanks to the liberality with which the means of conducting the experiment were placed at his disposal by Messrs Chance Brothers, of Birmingham, the question may perhaps be considered settled. After some preliminary trials, a pot of glass free from striæ was prepared of titanate of potash mixed with the ordinary ingredients of a crown glass. As the object of the experiment was merely to determine, in the first instance, whether titanic acid did or did not confer on the glass the unusual property of separating the colours at the blue end of the spectrum materially more, and at the red end materially less, than corresponds to a similar dispersive power in ordinary glasses, it was not thought necessary to employ pure titanic acid; and rutile fused with carbonate of potash was used as titanate of potash. The glass contained about 7 per cent. of rutile; and as rutile is mainly titanic acid, and none was

* *Report* for 1874, Transactions of the Sections, p. 26.

lost, the percentage of titanic acid cannot have been much less. The glass was naturally greenish, from iron contained in the rutile; but this did not affect the observations, and the quantity of iron would be too minute sensibly to affect the irrationality.

Out of this glass two prisms were cut. One of these was examined as to irrationality by Professor Stokes, by his method of compensating prisms, the other by Mr Hopkinson, by accurate measures of the refractive indices for several definite points in the spectrum. These two perfectly distinct methods led to the same result—namely, that the glass spaces out the more as compared with the less refrangible part of the spectrum no more than an ordinary glass of similar dispersive power. As in the phosphatic series, the titanium reveals its presence by a considerable increase of dispersive power; but, unlike what was observed in that series, it produces no sensible effect on the irrationality. The hopes, therefore, that had been entertained of its utility in silicic glasses prepared for optical purposes appear doomed to disappointment.

P.S.—Mr Augustus Vernon Harcourt has now completed an analytical determination which he kindly undertook of the titanic acid. From 2·171 grammes of the glass he obtained ·13 gramme of pure titanic acid, which is as nearly as possible 6 per cent.

44.

CERTAIN CASES OF ELECTROMOTIVE FORCE SUS-
TAINED BY THE ACTION OF ELECTROLYTES
ON ELECTROLYTES.

[From the *Proceedings of the Royal Society*, No. 166, 1876.]

In the following experiments the electromotive force was observed by a quadrant electrometer arranged for maximum sensibility; the connexions were made through the reversing-key; and, excepting the time observations for polarization, the readings were made twice at least on each side of the zero-point. A single Daniell's element gave 105 divisions deflection each way, so that in the following the unit of electromotive force is $\frac{1}{105}$ the electromotive force of Daniell's element. In all cases the electrodes were platinum wires dipping into the fluid under examination.

In the experiments on polarization the circuit was readily closed for a specified time by bringing the platinum wires into contact, and broken by releasing them; the electromotive force could then be observed at any instant after breaking the circuit.

I. Strong sulphuric acid was poured into a test-tube, which dipped into a porcelain crucible containing caustic potash. Thus the acid and alkali were separated by the glass of the tube. Platinum electrodes dipped into the two liquids. Electromotive force of 70 divisions was observed, the acid being positive. The crucible was heated by a spirit-lamp till the potash began to boil, the electromotive force increased to 153. The lamp was removed and the crucible allowed to cool; the electromotive force steadily diminished to 78 in half an hour. The tube was then discharged and insulated to observe the rate at which the charge developed.

E. F.

15 seconds after insulation, 67

30   „   „   69

60   „   „   69

II. The tube, crucible, and wire were thoroughly washed with cold water and replaced in position, but with water in place of both the acid and the alkali. The inside of the tube where the acid had been was now *negative*. E. F. = 16½. Heating to boiling the deflection increased to 150 divisions, but sank to 2 on cooling.

III. The test-tube contained potash and dipped into water in the crucible. When cold, E. F. = 33, the potash being negative ; when heated to boiling, E. F. = 36½.

IV. The tube contained strong sulphuric acid and dipped into water; a deflection of less than three divisions was observed. When heated till the water boiled, the reading was 35. After cooling the deflection decreased to 5.

V. The test-tube was removed and a small porcelain crucible introduced in its place ; sulphuric acid was poured into the outer crucible, potash into the inner ; platinum electrodes dipped into the liquids. On heating till the potash boiled, the electromotive force rose as high as 162. The decrease of the electromotive force as the liquids cooled was then observed.

Time in minutes.	E. F.	
0	155	boiling.
32	94	still warm.
91	88	quite cold.
181	88	

The author supposed these effects to be due to electrolytic action through the glass, not suspecting the true cause, excepting in V. But Sir William Thomson pointed out to him that the rate of development of the charge was greater than could occur *through* a substance of the low conductivity of the most conducting glass, and that the circuit must have been completed by conduction through a film of moisture on the surface of the glass. The next two experiments prove this to be the case.

VI. Drops of strong sulphuric acid and of caustic potash were placed on a sheet of common window-glass, previously carefully cleaned, but exposed to the air of the room. Platinum electrodes, dipping into each drop, communicated with the electrometer.

1. Drops half an inch apart, E. F. = 47.

2. The drops were connected by a thin trail of alkali drawn from the drop of alkali; E. F. = 105.

3. Drops 5 inches apart, two minutes being allowed for the charge to develop; E. F. = 12.

4. A trail was drawn halfway from one drop to the other; E. F. = 31.

5. The trail of liquid was continued till but $\frac{1}{4}$ inch of clear glass separated the liquids; E. F. = 43.

6. The connecting trail was completed from one drop to the other; E. F. = 70. It was observed that the potash trail had dried up, leaving a line of alkali between the drops.

VII. A dry chip of deal 6 inches long was split at each end, and a platinum wire let into each slit; the two wires were moistened with sulphuric acid and potash respectively at the points of contact with the wood; E. F. = 43.

VIII. Clean platinum wires were let into slits in a second dry chip of deal 12 inches long; these were connected for twenty minutes with the poles of a battery of two Daniell's elements, and then detached and connected through the reversing-key with the electrometer. As was expected an electromotive force opposite to that of the battery was observed, at first amounting to 33 divisions.

These experiments show that imperfect insulation, such as glass exposed to the air or wood, may cause errors in electrical experiments, not merely by leakage, but by introducing unknown electromotive forces, arising either from the imperfect insulators connecting different liquids, or from electrolytic polarization after a current has for some time been creeping through or over the surface of the insulators.

Several experiments were then tried on the direct action of liquids on liquids; two only are given here, because determinations have been made by other methods by Becquerel and others.

IX.   In a previous experiment a plug of moist sand had been rammed into the bend of a U-tube, and strong sulphuric acid and caustic potash poured into the limbs.   When this tube was washed out, it was found that a plug of sulphate of potash and sand $\frac{3}{16}$ inch thick had formed across the middle of the bend.   Strong sulphuric acid and potash were again poured into the limbs, and were now only separated by a thin plug of sulphate of potash. Platinum electrodes dipped into the liquids.   Electromotive force 139.   The circuit was closed for ten minutes.

			E. F.
14 seconds after insulation,			31
20	,,	,,	39
40	,,	,,	44
1 minute		,,	48
3 minutes		,,	74
5	,,	,,	89
10	,,	,,	98
20	,,	,,	105
65	,,	,,	119

The wires were again connected and the circuit left closed for about twenty-four hours.   It was found that the plug had extended for about half an inch on the side of the sulphuric acid by the formation of crystals of sulphate of potash, but had not apparently changed where it was in contact with the potash.

X.   A similar plug was formed in a second tube.   Into one limb sulphuric acid, with a small quantity of permanganate of potash, was poured, into the other caustic potash : E. F. = 178. Circuit was closed for ten minutes.

			E. F.
10 seconds after insulation,			90
20	,,	,,	110
40	,,	,,	123
1 minute		,,	128
3 minutes		,,	138
9	,,	,,	148
19	,,	,,	150

45.

ON THE QUASI-RIGIDITY OF A RAPIDLY MOVING CHAIN.

[From the *Proceedings of the Birmingham Philosophical Society.*]

Read May 9, 1878.

As diagrams would be necessary to an intelligible description of the apparatus employed or of the detail of the phenomena exhibited, it appears well to confine this abstract to a statement and short explanation of the more general dynamical properties of a moving chain, the more so as the experiments are very fully described by the inventor of most of them, Mr Aitkin.

Briefly, the apparatus consists of an endless chain hanging in a loop over a pulley which could be caused to revolve about a horizontal axis, so giving a rapid motion to the chain. It is firstly observed that the motion of the chain does not very materially affect the form in which the chain hangs when it attains equilibrium or a state of steady motion. The chain being at rest its form is a catenary : what forces must be applied to each small portion of the chain to keep the form the same when it is in motion ? Any such small portion is at any point moving, with velocity (V) the same for all points of the chain; hence, if R be the radius of the circle most nearly agreeing with the chain at the point (the circle of curvature), it follows that the change of motion is towards the centre of this circle at a rate $\dfrac{V^2}{R}$. Now it is also easy to show that a tension (T) in the chain will give a resolved force towards the centre on an element of chain, length ds, mass mds, equal to $\dfrac{Tds}{R}$. If, then, the tension of the chain be increased beyond that due to the forces acting upon it when at rest by the amount mV^2, constant for all parts of the chain and quite independent of R, this will be precisely sufficient to effect the actual changes of motion when the velocity is V, and the effect of such velocity will be not to alter the form but merely the tension of the chain.

Consider now a straight chain, stretched with tension T. Let the chain be struck at any point; two waves will be caused travelling in opposite directions with velocity $\sqrt{\dfrac{T}{m}}$. The height of these waves will be greater as the blow is greater, and less as the tension is greater; in fact, the height of the wave will vary directly as the blow, and inversely as the velocity of transmission of a wave. Suppose, now, the stretched chain be caused to move along its length with velocity V, $=\sqrt{\dfrac{T}{m}}$, how will these waves appear? That which is moving in a direction opposite to the motion of the chain will appear stationary to the observer as a rumple on the chain, whilst the other will appear to move away with velocity $2V$. It will also appear to such observer that to produce a rumple of given height he must strike a greater blow as the velocity and tension of the chain are greater: that is, if the velocity of the chain be doubled he must either strike twice as hard or strike two blows of the same value; or, if he be applying a continuous force to raise the rumple, he must apply it either twice as hard or twice as long. Let now the moving chain be curved, not straight; any small length of it may be regarded as sensibly straight, and we may conclude that the effect of any very small blow will be the same as if all the chain were straight, thus far, that it will cause a rumple fixed relatively to the observer, of which the height is inversely proportional to the velocity V, and a wave which will run away at a velocity $2V$.

We may now further explain the observations. When the chain is hanging in a catenary and in rapid motion, strike it a blow. As we should expect from the foregoing reasoning, the effect is different on a moving chain and on one at rest.   The chain presents a sort of rigidity greater as the velocity is greater; the blow causes a rumple or dint, which would remain firm in position but for the action of gravity.   Suppose the blow to be struck at the ascending side of the loop, two effects are observed. The rumple just mentioned travels downward with decreasing velocity till it reaches the bottom of the loop, where it remains as an almost permanent deformation; but besides this, sensibly at the instant when the blow is struck, a second rumple appears on the chain at the point where it meets the pulley, and travels downwards like the first with continually diminishing velocity. The explanation is easy:—As already shown the tension of the chain at any point consists of two parts, that due to the weight of the chain below and that due to the velocity of the chain,—the velocity of the chain is then less than that corresponding to the tension.   If we strike the chain we shall have two waves produced, one not quite stationary, but travelling slowly in a direction opposite to the motion of the chain, and stopping when it reaches the bottom of the loop, where the tension of the chain at rest is sensibly nil; the other, running up with a velocity a little more than double that of the chain, is reflected at the pulley, and then travels slowly downwards like the first.

The above will suggest the explanation of many other experiments.   We will here only deal with one as a further example. The chain is kept in contact with one point of the pulley by means of a second pulley, pressed by the hand against it in a horizontal direction at the point where it comes in contact with the first pulley on the ascending side; a piece of board is brought into contact with the lowest point of the loop of chain and somewhat rapidly raised—the chain stands up upon the board like a hoop of wire, rising up from the pulley to a height of perhaps three or four feet above it.   The pressure of the board in the first instance diminishes the tension of the chain at its lowest point.   This diminution will instantly extend throughout the chain, and may render the tension even at the highest point of the chain less than that due to the velocity.   If that be so, that highest point will recede from the centre about which it is moving—that is, will rise from the pulley.

46.

ON THE TORSIONAL STRAIN WHICH REMAINS IN A GLASS FIBRE AFTER RELEASE FROM TWISTING STRESS.

[From the *Proceedings of the Royal Society*, No. 191, 1878.]

Received October 4, 1878.

IT has long been known that if a wire of metal or fibre of glass be for a time twisted, and be then released, it will not at once return to its initial position, but will exhibit a gradually decreasing torsion in the direction of the impressed twist. The subject has undergone a good deal of investigation, especially in Germany. The best method of approximating to an expression of the facts has been given by Boltzmann (*Akad. der Wissensch. Wien*, 1874). He rests his theory upon the assumption that a stress acting for a short time will leave after it has ceased a strain which decreases in amount as time elapses, and that the principle of superposition is applicable to these strains, that is to say, that we may add the after-effects of stresses, whether simultaneous or successive. Boltzmann also finds that, if $\phi(t)\tau$ be the strain at time t resulting from a twist lasting a very short time τ, at time $t = 0$,

$$\phi(t) = \frac{A}{t},$$ where A is constant for moderate values of t, but

decreases when t is very large or very small. A year ago I made a few experiments on a glass fibre which showed a deviation from Boltzmann's law. A paper on this subject by Kohlrausch (*Pogg. Ann.*, 1876) suggested using the results of these experiments to examine how Boltzmann's law must be modified to express them. Professor Kohlrausch's results indicate that in the cases of silver wire and of fibre of caoutchouc Boltzmann's principle of super-position is only approximate, and that in the case of a short duration of twisting $\phi(t) = \dfrac{A}{t^a}$, where α is less than unity; in case of a long duration of twisting he uses other formulæ, which pretty successfully express his results, owing in part no doubt to the fact that in most cases each determination of the constants applies only to the results of one duration of twisting. In a case like the present it appears best to adopt a simple form involving constants for the *material* only, and then see in what way it fails to express the varying conditions of experiment. In 1865 Sir W. Thomson published (*Proceedings of the Royal Society*) the results of some experiments on the viscosity of metals, the method being to determine the rate at which the amplitude of torsional vibrations subsided. One of the results was that if the wire were kept vibrating for some time it exhibited much greater viscosity than when it had long been quiescent. This should guard us from expecting to attain great uniformity in experiments so roughly conducted as those of the present paper.

2. The glass fibre examined was about 20 inches in length. Its diameter, which might vary somewhat from point to point, was not measured. The glass from which it was drawn was composed of silica, soda, and lime; in fact, was glass No. 1 of my paper on "Residual Charge of the Leyden Jar" (*Phil. Trans.*, 1877). In all cases the twist given was one complete revolution. The deflection at any time was determined by the position on a scale of the image of a wire before a lamp, formed by reflection from a light concave mirror, as in Sir W. Thomson's galvanometers and quad-rant electrometer. The extremities of the fibre were held in clamps of cork; in the first attempts the upper clamp was not disturbed during the experiment, and the upper extremity of the fibre was assumed to be fixed; the mirror also was attached to the lower clamp. This arrangement was unsatisfactory, as one could not be certain that a part of the observed after-effect was not due to the fibre twisting within the clamps and then sticking. The difficulty was easily avoided by employing two mirrors, each cemented at a single point to the glass fibre itself, one just below the upper clamp, the other just above the lower clamp. The

upper mirror merely served by means of a subsidiary lamp and scale to bring back the part of the fibre to which it was attached to its initial position. The motion of the lower clamp was damped by attaching to it a vane dipping into a vessel of oil. The temperature of the room when the experiments were tried ranged from 13° C. to 13·8° C., and for the present purpose may be regarded as constant. The lower or reading scale had forty divisions to the inch, and was distant from the glass fibre and mirror 38¾ inches, excepting in Experiment V, when it was at 37½ inches. Sufficient time elapsed between the experiments to allow all sign of change due to after-effect of torsion to disappear. In all cases the first line of the table gives the time in minutes from release from torsion, the second the deflection of the image from its initial position in scale divisions.

Experiment I.—The twisting lasted 1 minute.

t	1	2	3	4	5	7	10	17	25
Scale divisions...	22	13	9	7	5½	4	3	2	1

Experiment II.—The twisting lasted 2 minutes.

t	1	2	3	4	5	7	10	20	40
Scale divisions...	38	25	18	15	13	10	8	4½	3½

Experiment III.—Twisted for 5 minutes.

t	1	2	3	4	5	7
Scale divisions...	64	51	41½	35½	32	26½

t	10	15	22	58	15
Scale divisions...	21½	17	14	7	2

Experiment IV.—Twisted for 10 minutes.

t	½	1	2	3	4	7	10
Scale divisions...	106	85	66	57	49½	37½	31

t	15	25	45	120	170
Scale divisions...	24½	18	13	7	6

Experiment V.—Twisted for 20 minutes.

t	1	2	3	4	5	7	10
Scale divisions...	110	89	75	68	61½	52	44

t	15	25	40	60	80	100
Scale divisions...	35½	26½	21	18	13½	12½

Experiment VI.—Twisted for 121 minutes.

t	$\frac{1}{2}$	1	2	3	4	5	7
Scale divisions...	191	170	148	136	$126\frac{1}{2}$	$119\frac{1}{2}$	$108\frac{1}{2}$

t	10	15	30	65	90	120	589
Scale divisions...	97	$84\frac{1}{2}$	$63\frac{1}{2}$	$41\frac{1}{2}$	34	28	$3\frac{1}{2}$

It should be mentioned that the operations of putting on the twist and of releasing each occupied about two seconds, and were performed half in the second before the beginning and end respectively of the period of twisting, and half in the second after or as nearly so as could be managed. The time was taken by ear from a clock beating seconds very distinctly.

3. The first point to be ascertained from these results is whether or not the principle of superposition, assumed by Boltzmann, holds for torsions of the magnitude here used.

If the fibre be twisted for time T through angle X, then the torsion at time t after release will be $X\{\psi(T+t) - \psi(t)\}$, where

$$\psi(t) = \int\phi(t)\,dt.$$

If now $T = t_1 + t_2 + t_3 + \dots$ we may express the effect of one long twist in terms of several shorter twists by simply noticing that

$$
\begin{aligned}
X\{\psi(t) - \psi(t+T)\} = X\,[& \{\psi(t) - \psi(t+t_1)\} \\
& + \{\psi(t+t_1) - \psi(t+t_1+t_2)\} \\
& + \{\psi(t+t_1+t_2) - \psi(t+t_1+t_2+t_3)\} + \&c.].
\end{aligned}
$$

Apply this to the preceding results, calculating each experiment from its predecessor. Let x_t be the value of $\psi(T+t) - \psi(t)$, that is, the torsion at time t, when free, divided by the impressed twist measured in same unit; we obtain the following five tables of comparison.

Results for $T = 2$ compared with those from $T = 1$.

t	1	2	3	4	5	7
x_t observed	0·00195	128	092	077	066	051
x_t calculated	0·00199	112	082	064	051	040

t	10	20	40
x_t observed	041	023	018
x_t calculated	029	016	

Results for $T = 5$ compared with those from $T = 2$ and $T = 1$.

t	1	2	3	4	5	7	10
x_t observed	0·00328	262	212	182	164	136	110
x_t calculated	0·00323	233	181	156	136	108	193

t	15	22	58	151
x_t observed	087	072	036	010
x_t calculated	066	047		

Results for $T = 10$ compared with those from $T = 5$.

t	$\frac{1}{2}$	1	2	3	4	7	10
x_t observed	0·00544	435	338	292	253	192	159
x_t calculated	—	469	398	339	300	236	197

t	15	25	45	120	170
x_t observed	125	092	067	036	031
x_t calculated	161	130	088		

Results for $T = 20$ compared with those from $T = 10$.

t	1	2	3	4	5	7	10
x_t observed	0·00580	470	398	358	327	276	234
x_t calculated	0·00587	483	430	384	356	312	266

t	15	25	40	60	80	100
x_t observed	188	140	111	085	072	066
x_t calculated	217	167	135	100	084	

Results for $T = 121$ compared with those from $T = 20$.

t	$\frac{1}{2}$	1	2	3	4	5	7
x_t observed	0·00979	871	758	697	648	612	556
x_t calculated	—	1070	950	880	830	780	730

t	10	15	30	65	90	120	589
x_t observed	497	433	325	212	174	144	18
x_t calculated	670	600	500	380	350		

In examining these results it must be remembered that those for small values of T are much less accurate than when T is greater, for the quantity observed is smaller but is subject to the same absolute error; any irregularity in putting on or releasing from the stress will cause an error which is a material proportion of the observed deflection. For this reason it would be unsafe to base a conclusion on the experiments with $T=1$ and $T=2$. The three last tables agree in indicating a large deviation from the principle of superposition, the actual effect being *less* than the sum of the separate effects of the periods of stress into which the actual period may be broken up. Kohlrausch finds the same to be the case for india-rubber, either greater torsions or longer durations give less after-effects than would be expected from smaller torsions and shorter periods.

4. Assuming with Boltzmann that $\phi(t) = \dfrac{A}{t}$, we have at time t after termination of a twist lasting time T,

$$x_t = A \{\log(T+t) - \log t\},$$

the logarithms being taken to any base we please. The results were plotted on paper, x_t being the ordinate and $\log \dfrac{T+t}{t}$ the abscissa; if the law be true we should find the points all lying on a straight line through the origin. For each value for T they do lie on straight lines very nearly for moderate values of t; but if T is not small these lines pass above the origin. When t becomes large the points drop below the straight line in a curve making towards the origin. This deviation appears to indicate the form $\phi(t) = \dfrac{A}{t^a}$, α being less than, but near to, unity. If $\alpha = 0.95$ we have a fairly satisfactory formula:

$$x_t = A'(\overline{T+t}^{\frac{1}{20}} - t^{\frac{1}{20}}), \text{ where } A' = \frac{A}{1-\alpha} \text{ when } T = 121.$$

In the following Table the observed and calculated values of x_t when $T = 121$ are compared, A' being taken as 0·032.

t	$\tfrac{1}{2}$	1	2	3	4	5	7
x_t observed	0·00979	871	758	697	648	612	556
x_t calculated	0·00976	870	755	691	643	600	550

t	10	15	30	65	90	120	589
x_t observed	497	433	325	212	174	144	18
x_t calculated	493	429	320	218	176	147	42

To show the fact that A' decreases as T increases if α be assumed constant, I add a comparison when $T = 20$, it being then necessary to take $A' = 0.037$.

t	1	2	3	4	5	7	10
x_t observed	0.00580	470	398	358	327	276	234
x_t calculated	0.00607	485	422	370	337	285	233

t	15	25	40	60	80	100
x_t observed	188	140	111	085	072	066
x_t calculated	185	125	089	067	052	041

A better result would in this case be obtained by assuming $\alpha = 0.92$, or $= 0.93$ in the former case with $A' = 0.021$. Probably the best result would be given by taking A constant, and assuming that α increases with T.

Taking the formula $\phi(t) = \dfrac{A}{t}$ these experiments give values of A ranging from 0.0017 to 0.0022. Boltzmann for a fibre, probably of a quite different composition, gives numbers from which it follows that $A = 0.0036$.

5. In my paper on "Residual Charge of the Leyden Jar*" that subject is discussed in the same manner as Boltzmann discusses the after-effect of torsion on a fibre, and it is worth remarking that the results of my experiments can be roughly expressed by a formula in which $\phi(t) = \dfrac{A}{t^a}$. For glass No. 5 (soft crown) $\alpha = 0.65$, whilst for No. 7 (light flint) it is greater; but in the electrical experiment no sign of a definite deviation from the law of super-position was detected.

* *Supra*, p. 19.

47.

ON THE STRESSES CAUSED IN AN ELASTIC SOLID BY INEQUALITIES OF TEMPERATURE*.

[From the *Messenger of Mathematics*, New Series, No. 95, March, 1879.]

VARIOUS phenomena due to the stresses caused by inequalities of temperature will occur to everyone. Glass vessels crack when they are suddenly and unequally heated, or when in manufacture they have been allowed to cool so as to be in a state of stress when cold. Optical glass is doubly refracting when badly annealed or when different parts of the mass are at different temperatures. Iron castings which have been withdrawn from the mould whilst still very hot, or of which the form is such that some parts cool more rapidly than others, are liable to break without the application of any considerable external stress. The ordinary theory of elastic solids may easily be applied to some such cases.

Let u, v, w be the displacements of any point (xyz) of a body density ρ, parallel to the coordinate axes. Let N_1, N_2, N_3, T_1, T_2, T_3 be the elements of stress; i.e. $N_1\alpha$ is the tension across an elementary area α resolved parallel to x, the element α being perpendicular to x; $T_1\beta$ is the shearing force across an element β resolved parallel to z, β being perpendicular to y; T_1 is then also the shear parallel to y across an element perpendicular to z.

---

* See *Report of the British Association* for 1872, p. 51.

If $\rho X, \rho Y, \rho Z$ be the external forces at (xyz)

$$
\left.
\begin{aligned}
\frac{dN_1}{dx} + \frac{dT_3}{dy} + \frac{dT_2}{dz} + \rho X = 0 \\[2mm]
\frac{dT_3}{dx} + \frac{dN_2}{dy} + \frac{dT_1}{dz} + \rho Y = 0 \\[2mm]
\frac{dT_2}{dx} + \frac{dT_1}{dy} + \frac{dN_3}{dz} + \rho Z = 0
\end{aligned}
\right\} \quad \dots\dots\dots\dots(1).
$$

These are strictly accurate. Of an inferior order of accuracy are the equations expressing the stresses in terms of the strains of an isotropic solid

$$
\left.
\begin{aligned}
N_1 = \lambda\theta + 2\mu\,\frac{du}{dx} \\[2mm]
T_1 = \mu\left(\frac{dv}{dz} + \frac{dw}{dy}\right)
\end{aligned}
\right\} \quad \dots\dots\dots\dots\dots(2),
$$

where $\theta = \dfrac{du}{dx} + \dfrac{dv}{dy} + \dfrac{dw}{dz} =$ the dilatation at the point. These equations are inaccurate, inasmuch as they are inapplicable if the strains be not very small, and as even then in all solids which have been examined the stresses depend not only on the then existing strains but in some degree on the strains which the body has suffered in all preceding time (see Boltzmann, *Akad. der Wissensch. zu Wien*, 1874; Kohlrausch, *Pogg. Annalen*, 1876; Thomson, *Proceedings of Royal Society*, 1865; some experiments of my own, *Proceedings of Royal Society*, 1878*; Viscosity in Maxwell's *Heat*).

Assuming equations (2) we observe that as these and also (1) are linear, we may superpose the effects of separate causes of stress in a solid when they act simultaneously.

Equations (2) are intended to apply only to cases in which when the stresses vanish the strains vanish, and in which the strains result from stress only and not from inequalities of temperature. The first limitation is easily removed by the principle of superposition. We must determine separately the stresses when no external forces are applied, and then the stresses due to the external forces on the assumption that the solid is unstrained when free and finally add the results. For example, if we are considering a gun or press cylinder, we know that internal

---

* *Supra*, p. 350.

pressure will produce the greatest tension in the inner shells, and we can hence at once infer that if the gun or cylinder be so made that normally the inner shells are in compression and the outer in tension it will be stronger to resist internal pressure.

To ascertain the effect of unequal heating, assume that λ, μ are independent of the temperature, an assumption of the same order of accuracy as assuming in the theory of conduction of heat that the conductivity is a constant independent of the temperature. Let κ be the coefficient of linear expansion, τ the temperature at any point in excess of a standard temperature. If there be no stresses,

$$\frac{du}{dx} = \frac{dv}{dy} = \frac{dw}{dz} = \kappa\tau \; ;$$

therefore

$$(3\lambda + 2\mu)\,\kappa\tau = \lambda\theta + 2\mu\,\frac{du}{dx},$$

$$0 = \mu\left(\frac{dw}{dy} + \frac{dv}{dz}\right), \text{ &c.} \; ;$$

if there were stresses, but τ were zero,

$$N_1 = \lambda\theta + 2\mu\,\frac{du}{dx}, \text{ &c.} \; ;$$

superposing effects we have

$$\left.\begin{aligned} N_1 &= \lambda\theta + 2\mu\,\frac{du}{dx} - (3\lambda + 2\mu)\,\kappa\tau \\[2mm] T_1 &= \mu\left(\frac{dw}{dy} + \frac{dv}{dz}\right) \end{aligned}\right\} \quad \dots\dots\dots(3).$$

Substitute in the equations of equilibrium

$$\left.\begin{aligned} (\lambda + \mu)\,\frac{d\theta}{dx} + \mu\nabla^2 u - \gamma\frac{d\tau}{dx} + \rho X &= 0 \\[2mm] (\lambda + \mu)\,\frac{d\theta}{dy} + \mu\nabla^2 v - \gamma\frac{d\tau}{dy} + \rho Y &= 0 \\[2mm] (\lambda + \mu)\,\frac{d\theta}{dz} + \mu\nabla^2 w - \gamma\frac{d\tau}{dz} + \rho Z &= 0 \end{aligned}\right\} \dots\dots\dots(4),$$

where

$$\gamma = (3\lambda + 2\mu)\,\kappa.$$

If there be equilibrium of temperature $\nabla^2\tau = 0$, *and the effect of unequal heating is exactly the same as that of an external force potential* $\dfrac{(3\lambda + 2\mu)\,\kappa\tau}{\rho}$; in this case we have the equations

$$\left.\begin{aligned}\nabla^2\theta &= \rho\,\frac{d^2\theta}{dt^2}\\[2mm]\nabla^2\nabla^2 u &= 0\end{aligned}\right\}\dotfill(5),$$

still true and under the same conditions.

Examine the case when there are no bodily forces and when everything is symmetrical about a centre. The displacement at any point is radial, call it U, and the principal stresses are radial and tangential, call them R and T.

The equation of equilibrium is

$$\frac{d(r^2 R)}{dr} - 2rT = 0 \dotfill(6),$$

and the stresses are expressed by

$$\left.\begin{aligned}R &= \lambda\theta + 2\mu\,\frac{dU}{dr} - \gamma\tau\\[2mm]T &= \lambda\theta + 2\mu\,\frac{U}{r} - \gamma\tau\\[2mm]\theta &= \frac{1}{r^2}\,\frac{d(r^2 U)}{dr}\end{aligned}\right\}\dotfill(7);$$

substituting

$$(\lambda + 2\mu)\left\{\frac{d^2 U}{dr^2} + 2\,\frac{1}{r}\,\frac{dU}{dr} - 2\,\frac{U}{r^2}\right\} = \gamma\,\frac{d\tau}{dr}\dotfill(8),$$

therefore

$$r^2 U = \frac{\gamma}{\lambda + 2\mu}\int r^2\tau\,dr + ar^3 + b \dotfill(9),$$

where a and b are constants to be determined by a knowledge of R or U for two specified values of r. This equation is of course true whether there be equilibrium of temperature or not.

The interior and exterior surfaces of a homogeneous spherical shell are maintained at different temperatures, to find the resulting stresses.

Let r_1, r_2 be the internal and external radii, t_1, t_2 the internal and external temperatures, then if τ be the temperature at radius r,

$$\tau = c + \frac{f}{r}$$

where

$$c = \frac{r_2 t_2 - r_1 t_1}{r_2 - r_1}$$

and

$$f = -\frac{(t_2 - t_1)\, r_1 r_2}{r_2 - r_1}$$

$$\left. \right\} \quad \dots\dots\dots\dots\dots(10).$$

Substitute in equation (9) and then in (7)

$$U = \frac{\gamma}{\lambda + 2\mu} \left(\tfrac{1}{3}cr + \tfrac{1}{2}f\right) + ar + \frac{b}{r^2}$$

$$\theta = \frac{\gamma}{\lambda + 2\mu}\left(c + \frac{f}{r}\right) + 3a = \frac{\gamma}{\lambda + 2\mu}\,\tau + 3a$$

$$R = -\frac{4\mu\gamma}{3(\lambda + 2\mu)}c - \frac{2\mu\gamma}{\lambda + 2\mu}\frac{f}{r} + (3\lambda + 2\mu)\,a - \frac{4\mu b}{r^3}$$

$$\left. \right\} \dots(11),$$

write R in the form

$$R = A + \frac{B}{r} + \frac{C}{r^3};$$

where

$$B = \frac{2\mu\gamma}{\lambda + 2\mu}\frac{(t_2 - t_1)\, r_1 r_2}{r_2 - r_1},$$

for we shall not require to find U; we find

$$A(r_2^3 - r_1^3) + B(r_2^2 - r_1^2) = 0,$$

$$C(r_2^3 - r_1^3) + B r_1^2 r_2^2 (r_2 - r_1) = 0,$$

whence

$$R = \frac{2\mu\gamma (t_2 - t_1)\, r_1 r_2}{r_2^3 - r_1^3}\left\{-(r_2 + r_1) + \frac{r_2^2 + r_1 r_2 + r_1^2}{r} - \frac{r_1^2 r_2^2}{r^3}\right\}\dots\dots(12).$$

R will have a maximum or minimum value when

$$r^2 = \frac{3 r_1^2 r_2^2}{r_2^2 + r_1 r_2 + r_1^2},$$

and its value then is

$$B'\left\{-(r_2 + r_1) + \frac{2(r_2^2 + r_1 r_2 + r_1^2)^{\frac{3}{2}}}{3 r_1 r_2 \sqrt{3}}\right\};$$

this is positive if $t_2 > t_1$, as we may see at once from physical reasons.

Now

$$T = \frac{1}{2r}\frac{dr^2R}{dr} = B'\left\{-(r_2 + r_1) + \frac{r_2^2 + r_1r_2 + r_1^2}{2r} + \frac{r_1^2r_2^2}{2r^3}\right\},$$

if $t_2 > t_1$; this decreases as r increases; when $r = r_1$, its value becomes

$$B'\left\{\frac{-2(r_2 + r_1)r_1 + (r_2^2 + r_1r_2 + r_1^2) + r_2^2}{2r_1}\right\}$$

$$= \frac{\mu\gamma}{\lambda + 2\mu}\frac{(t_2 - t_1)r_2}{r_2^3 - r_1^3}(r_2 - r_1)(2r_2 + r_1) \ldots\ldots\ldots(13)^*.$$

The case when the thickness is small is interesting. Let $r_2 = r_1 + x$, then the maximum tension is

$$\frac{\mu\gamma(t_2 - t_1)}{(\lambda + 2\mu)}\left(1 + \frac{2x}{3r_1}\right);$$

neglecting the term $\dfrac{x}{r_1}$ in comparison with unity, we see that of two vessels the thicker is not sensibly more liable to break than the thinner, a result at first sight contradictory to experience. The explanation is that the greater liability of thick vessels to break is due to the fact that, allowing heat to pass through but slowly, a greater difference of temperature between the two surfaces really exists.

Let t_2', t_1' be the actual surface temperatures, we may assume that, if t_2 and t_1 be the temperatures of the surrounding media, the heat passing the two surfaces per unit of area will be $H_2(t_2 - t_2')$ and $H_1(t_1' - t_1)$.

Hence

$$t_2' - t_1' = \frac{H_1H_2(t_2 - t_1)}{(H_1 + H_2)\dfrac{\kappa}{x} + H_1H_2};$$

using this in the equation last obtained we have a result quite in accord with experience.

---

* This result was set by me in the recent Mathematical Tripos Examination (Friday afternoon, January 17, 1879, Question ix).

Returning now to equations (7) and (9), suppose the sphere to be solid and to be heated in any manner symmetrical about the centre. The constant b must vanish, and

$$\theta = \frac{\gamma}{\lambda + 2\mu} \, \tau + 3a,$$

$$R = (\lambda + 2\mu)\, \theta - 4\mu \frac{U}{r} - \gamma\tau$$

$$= -4\mu \frac{U}{r} + 3(\lambda + 2\mu)\, a$$

$$= (3\lambda + 2\mu)\, a - \frac{4\mu\gamma}{(\lambda + 2\mu)\, r^3} \int r^2 \tau dr.$$

Now the mean temperature within the radius r is

$$\frac{4\pi \int r^2 \tau dr}{\tfrac{4}{3}\pi r^3} = \frac{3 \int r^2 \tau dr}{r^3} \, ;$$

therefore, since the pressure is zero at the surface of the sphere,

$$R = \frac{4\mu\gamma}{3(\lambda + 2\mu)} \cdot \{\text{mean temperature of whole sphere} - \text{mean tem-}$$

perature of sphere of radius r}............(14),

$$T = \tfrac{1}{2} \frac{1}{r} \frac{dr^2 R}{dr} ....(15),$$

$$= \frac{4\mu\gamma}{3(\lambda + 2\mu)} \cdot \{\text{mean temperature of whole sphere} - \tfrac{3}{2}\tau + \tfrac{1}{2} \text{ mean}$$

temperature within the sphere of radius r}.........(16).

Other problems of the same character as the preceding will suggest themselves, for example that of a cylinder heated symmetrically around an axis, but as no present use could be made of the results I do not discuss them.

48.

ON THE THERMO-ELASTIC PROPERTIES OF SOLIDS.

(Published in 1879 as an Appendix to Clausius' "Theory of Heat*.")

Sir William Thomson was the first who examined the thermo-elastic properties of elastic solids. Instead of abstracting his investigation (*Quarterly Mathematical Journal*, 1855) it may be well to present the subject as an illustration of the method of treatment by the Adiabatic Function.

Consider any homogeneously-strained elastic solid. To define the state of the body as to strain six quantities must be specified, say u, v, w, x, y, z: these are generally the extensions along three rectangular axes, and the shearing strains about them, each relative to a defined standard temperature and a state when the body is free from stress. The work done by external forces when the strains change by small variations may always be expressed in the form

$$(U\delta u + V\delta v + ...) \times \text{volume of the solid,}$$

because the conditions of strain are homogeneous. U, V...... are the stresses in the solid: each is a function of u, v...... and of the temperature, and is determined when these are known. Let θ denote the temperature (where θ is to be regarded merely as the *name* of a temperature, and the question of how temperatures are to be measured is not prejudged).

Amongst other conditions under which the strains of the body may be varied, there are two which we must consider. First, suppose that the temperature is maintained constant; or that the

* Translation by W. R. Browne, M.A.; Macmillan and Co., p. 363.

change is effected *isothermally*. Then θ is constant. Secondly, suppose that the variation is effected under such conditions that no heat is allowed to pass into or to leave the body; or that the change is effected *adiabatically*. In the latter case θ, u, v, ... are connected by a relation involving a parameter which is always constant when heat does not pass into or out of the solid: this parameter is called the *adiabatic function*.

We have now fourteen quantities relating to the body, viz. six elements of strain, six of stress, the quantity θ which defines the temperature, and the parameter ϕ the constancy of which imposes the adiabatic condition. Any seven of these may be chosen as independent variables.

Let the body now undergo Carnot's four operations as follows:—

1°. Let the stresses and strains vary slightly under the sole condition that the temperature does not change. Let the consequent increase of ϕ be $\delta\phi$. Heat will be absorbed or given out, and, since the variations are small, the quantity will be proportional to $\delta\phi$, say

$$f(\theta, u, v, w, \ldots\ldots)\,\delta\phi.$$

2°. Let the stresses and strains further vary adiabatically, and let $\delta\theta$ be the consequent increase of temperature.

3°. Let the stresses and strains receive any isothermal variation, such that the parameter ϕ returns to its first value. Heat will be given out or absorbed, equal to

$$(f + \delta f)\,\delta\phi.$$

4°. Let the body return to its first state.

Here we have a complete and reversible cycle. The quantity of heat given off $\delta f \times \delta\phi$ is equal to the work done by external forces. Now Carnot's theorem (or the Second Principle of Thermodynamics) asserts that the work done, or $\delta f \times \delta\phi$, divided by the heat transferred from the lower to the higher temperature, $f \times \delta\phi$, is equal to a function of θ only (which function is the same for all bodies) multiplied by $\delta\theta$. Thus

$$\frac{\delta f}{f} = F(\theta)\,\delta\theta,$$

$$\log(f) = \int F(\theta)\,d\theta + \text{a function of } \phi;$$

the function of ϕ being added because the variation was performed under the condition that ϕ was constant. By properly choosing the parameter ϕ this function may be included in $\delta\phi$, and we have, as the quantity of heat absorbed in the first operation,

$$\epsilon^{\int F\theta d\theta} \times \delta\phi.$$

The mode of measuring temperature being arbitrary, we shall find it convenient to define that temperature is so measured that $F(\theta) = \dfrac{1}{\theta}$; then we have:—

Heat absorbed in first operation $= \theta\delta\phi$............(1);

Work done by external forces $\quad = \delta\theta \times \delta\phi$ ......(2).

We must now examine more particularly the variations in the stresses and strains. Denote the values of $U, V, \ldots\ldots, u, v, \ldots\ldots$ by different suffixes for the four operations.

The work done by the external forces in these operations is respectively

$$\frac{U_1 + U_2}{2}(u_2 - u_1) + \frac{V_1 + V_2}{2}(v_2 - v_1) + \&c.,$$

$$\frac{U_2 + U_3}{2}(u_3 - u_2) + \frac{V_2 + V_3}{2}(v_3 - v_2) + \&c.,$$

$$\frac{U_3 + U_4}{2}(u_4 - u_3) + \frac{V_3 + V_4}{2}(v_4 - v_3) + \&c.,$$

$$\frac{U_4 + U_1}{2}(u_1 - u_4) + \frac{V_4 + V_1}{2}(v_1 - v_4) + \&c.,$$

and the sum of these is equal to $\delta\phi\,\delta\theta$.

Hence a variety of important relations may be obtained.

Let all the strains but one be constant: then we have

$$u_2 = u_1 + \frac{du}{d\phi}d\phi,$$

$$u_3 = u_2 + \frac{du}{d\theta}d\theta,$$

$$u_4 = u_3 - \frac{du}{d\phi}d\phi,$$

$$u_1 = u_4 - \frac{du}{d\theta}d\theta,$$

with similar equations for U_2, &c.   Hence the Work done in the successive operations is,

$$\frac{U_1 + U_2}{2} \times \frac{du}{d\phi} \, d\phi,$$

$$\frac{U_2 + U_3}{2} \times \frac{du}{d\theta} \, d\theta,$$

$$-\frac{U_2 + \dfrac{dU}{d\theta} d\theta + U_1 + \dfrac{dU}{d\theta} d\theta}{2} \times \frac{du}{d\phi} \, d\phi,$$

$$-\frac{U_2 - \dfrac{dU}{d\phi} d\phi + U_3 - \dfrac{dU}{d\phi} d\phi}{2} \times \frac{du}{d\theta} \, d\theta.$$

Adding these, the total Work done becomes

$$\left(-\frac{du}{d\phi} \frac{dU}{d\theta} + \frac{du}{d\theta} \frac{dU}{d\phi} \right) \times d\theta d\phi,$$

the differentiations being performed when u and U are expressed as functions of θ, ϕ and the five other strains.

The same is true if five stresses are constant, that is if u and U are expressed as functions of θ, ϕ and the five other stresses.

But from (2) the Work done $= d\theta \times d\phi$.   Hence it follows generally (using the well-known theorem as to Jacobians) that

$$\frac{d\phi}{dU} \frac{d\theta}{du} - \frac{d\phi}{du} \frac{d\theta}{dU} = 1 \quad \dots\dots\dots\dots\dots(3),$$

ϕ and θ being expressed as functions of u, U; and either the five other stresses or the five other strains being constant.

These equations are still true if the independent variables are partly stresses and partly strains, so long as no two are of the same name : e.g. if they are $vwXYZ$.

From equation (3) all the thermo-elastic properties of bodies may be deduced.   We have generally

$$d\theta = \frac{d\theta}{dU} dU + \frac{d\theta}{du} du \dots\dots\dots\dots\dots(4),$$

$$d\phi = \frac{d\phi}{dU} dU + \frac{d\phi}{du} du \dots\dots\dots\dots\dots(5).$$

Putting $d\phi = 0$, we have

$$\frac{dU}{du} \text{ (when } \phi \text{ is constant)} = -\frac{\dfrac{d\phi}{du}}{\dfrac{d\phi}{dU}}.$$

Putting $d\theta = 0$, we have

$$\frac{dU}{du} \text{ (when } \theta \text{ is constant)} = -\frac{\dfrac{d\theta}{du}}{\dfrac{d\theta}{dU}}.$$

Let $\dfrac{d_\phi\theta}{du}$ denote $\dfrac{d\theta}{du}$ under the condition that ϕ is constant, that is, where θ is expressed as a function of ϕ, u instead of U, u. Then by (4)

$$\frac{d_\phi\theta}{du} = \frac{d\theta}{dU}\frac{dU}{du} + \frac{d\theta}{du} = -\frac{d\theta}{dU} \times \frac{\dfrac{d\phi}{du}}{\dfrac{d\phi}{dU}} + \frac{d\theta}{du}$$

$$= -\frac{1}{\dfrac{d\phi}{dU}}, \text{ by (3).}$$

This is the fourth thermodynamic relation (see Maxwell *on Heat*, 1877, p. 169).

The others are obtained in a similar way thus:—

$$\frac{d_\phi\theta}{dU} = \frac{d\theta}{dU} - \frac{d\theta}{du}\frac{\dfrac{d\phi}{dU}}{\dfrac{d\phi}{du}} = -\frac{1}{\dfrac{d\phi}{du}} = -\frac{d_U u}{d\phi},$$

$$\frac{d_\theta\phi}{du} = \frac{d\phi}{du} - \frac{d\phi}{dU}\frac{\dfrac{d\theta}{du}}{\dfrac{d\theta}{dU}} = -\frac{1}{\dfrac{d\theta}{dU}} = -\frac{d_u U}{d\theta},$$

$$\frac{d_\theta\phi}{dU} = \frac{d\phi}{dU} - \frac{d\phi}{du}\frac{dU}{\dfrac{d\theta}{du}} = \frac{1}{\dfrac{d\theta}{du}} = \frac{d_U u}{d\theta}.$$

These relations are true provided each of the other strains, or else its corresponding stress, is constant.

Take the last of these for interpretation.   When θ is constant we have by (1),

$$\text{Heat absorbed in any change (or } dq) = \theta d\phi.$$

Hence

$$\frac{d_\theta\phi}{dU} = \frac{1}{\theta}\frac{dq}{dU};$$

or, by the fourth relation,

$$\frac{d_v\phi}{d\theta} = \frac{1}{\theta}\frac{dq}{dU}.$$

Here $\dfrac{d_v u}{d\theta}$ is the coefficient of dilatation.   This, under the conditions assumed, will, of course, be different according as the other stresses or other strains are maintained constant.   In the case of a bar of india-rubber stretched by a variable weight, all the elements of stress but one vanish or are constant.   If the stress be somewhat considerable it is found that $\dfrac{du}{d\theta}$ is negative.   It follows that increase of weight will liberate heat in the india-rubber.   But the same will not be true if the stretching weight be nil or very small, nor again if the periphery of the bar is held so that it cannot contract transversely as the weight extends it longitudinally, unless (which is improbable) it should be found that in these cases the coefficient of dilatation is negative.

49.

ON HIGH ELECTRICAL RESISTANCES.

[From the *Philosophical Magazine*, March, 1879, pp. 162—164.]

In the *Philosophical Magazine* of July 1870 Mr Phillips describes a method of readily constructing very high electrical resistances. A pencil-line is ruled on glass; the ends of the line are provided with the means of making electrical connexion; and the whole is varnished: by this means a resistance of two million ohms was obtained; and it was found to be constant under varying potential. This method of constructing resistances is alluded to in Maxwell's *Electricity* (p. 392); but I do not know that it has received the examination it deserves, or that it has come into general use. Having need of resistances of over 100 million ohms, I have made a few on Mr Phillips's plan, ranging from 26,000 ohms to 96,000,000 ohms (which are fairly satisfactory), and one or two much greater (which do not conduct according to Ohm's law, but with a resistance diminishing as the electromotive force increases). A short description of these may perhaps save a little trouble to others who desire tolerably constant high resistances.

All my resistances are ruled on strips of patent plate glass which has been finished with fine emery, but has not been polished. The strips are twelve inches long, and, except in the cases specified below, about half an inch wide. One or more parallel lines are ruled on each strip, terminating at either end in a small area covered with graphite from the pencil. The strip of glass, first heated over a spirit-lamp, is varnished with shellac varnish, excepting only these small terminal areas, which are surrounded by a small cup of paraffin-wax to contain mercury to make the necessary connexions. To secure better insulation, feet of paraffin or of glass covered with paraffin are attached on the underside at the ends of the strip to support it from the table. Before varnishing, each strip was marked with a distinguishing letter. The strips marked g, h, i, a, and b were ruled with a BB pencil, the remainder with a HHH.

These resistances appear to be not quite constant, but to vary slightly with time, the maximum variation in four months being slightly in excess of $\frac{1}{2}$ per cent. In every case they were examined under varying potential to ascertain if they obeyed Ohm's law. With the exception of f, described below, all were satisfactory in this respect. The resistance appears to diminish slightly as the temperature rises; but this conclusion rests on a single rough experiment, and must be regarded as uncertain.

The values of the resistances were determined with a differential galvanometer, each coil having a resistance of 3500 ohms, by the well-known method of dividing a battery-current, passing one part through the large resistance to be measured and one coil of the galvanometer, the other through a set of coils or other known resistance, and then through the galvanometer shunted with a second set of resistance-coils. g was thus compared with standard coils. g was then used to find h and i; and $h + i$ was used to find a and b. A Thomson's quadrant electrometer was used to compare in succession k, l, and m with $a + b$. c and e were similarly compared with $k + l + m$; and, lastly, c and e were used to examine f.

g is ruled on a strip one inch wide, rather more than half the surface being covered with graphite. Three experiments on the same day gave 26,477, 26,461, and 26,470 ohms; the variations are probably due to uncertainty in the temperature-correction, the galvanometer-coils being of copper. After the lapse of four months 26,615 ohms was obtained.

i is ruled on a strip three-quarters of an inch wide, with nine tolerably strong lines; its resistance was first found to be 209,907 ohms, and four months later to be 208,840.

b has four strong lines on a strip half an inch wide; resistance 207,954 on a first occasion, and 208,750 after the lapse of four months.

a has two lines narrower than the preceding; resistance 5,240,000 at first, and 5,220,800 after four months.

h has a single line apparently similar to either of those of a; and the resistance is 9,168,000.

k, l, and m have each two lines ruled with a HHH pencil; their resistances are respectively 23,024,000, 14,400,000, and 13,218,000 ohms.

c and e also have two lines, but they are finer; the resistances are 79,407,000 and 96,270,000.

As already mentioned, all the preceding were tested with various battery-powers, and were found to obey Ohm's law within the limits of observation. It was not so with f, as the following observation shows very clearly. c, k, e, and f were arranged as a Wheatstone's bridge. Junctions (f, c) and (e, k) were connected to the poles of a Daniell's battery varying from one to eighteen elements; junctions (e, f) and (k, c) were respectively connected through the reversing-key with the quadrants of the electrometer. The potential of one Daniell's element was represented by 270 divisions of the scale of the electrometer. Column I. gives the number of elements employed, II. the corresponding reading of the electrometer, III. the value of $\dfrac{k}{k+c} - \dfrac{e}{f+e}$ deduced therefrom, and IV. the values of the ratio resistance of f : resistance of e.

I.	II.	III.	IV.
1	16	0·060	5·1
2	25	0·046	4·6
3	31½	0·039	4·4
4	31	0·029	4·1
5	28	0·021	3·9
6	27½	0·017	3·8
9	10	0·0041	3·5
12	− 5	− 0·0016	3·4
15	− 25	− 0·006	3·3
18	− 47	− 0·0097	3·25

This result is by no means surprising. There is doubtless an exceedingly minute discontinuity in the fine line across which disruptive discharge occurs; and the moral is, that resistances of this kind should always be tested as regards their behaviour under varying electromotive force.

Several attempts to rule a line on a strip 12 inches long with a resistance over 100,000,000 ohms resulted in failure.

50.

NOTE ON Mr E. H. HALL'S[*] EXPERIMENTS ON THE "ACTION OF MAGNETISM ON A PERMANENT ELECTRIC CURRENT."

[From the *Philosophical Magazine*, December, 1880, pp. 430, 431.]

If X, Y, Z be the components of electromotive force, and u, v, w the components of current at any point, in any body conducting electricity, we have the equations

$$\left.\begin{array}{l} X = R_1 u + S_3 v + S_2 w - Tv, \\ Y = S_3 u + R_2 v + S_1 w + Tu, \\ Z = S_2 u + S_1 v + R_3 w, \end{array}\right\}$$

where R_1, R_2, R_3, S_1, S_2, S_3, T are constants for the substance under its then circumstances (*vide* Maxwell's *Electricity*, vol. I. p. 349).

After obtaining these equations, Maxwell goes on to say:— "It appears from these equations that we may consider the electromotive force as the resultant of two forces, one of them depending on the coefficients R and S, and the other depending on T alone. The part depending on R and S is related to the current in the same way that the perpendicular on the tangent plane of an ellipsoid is related to the radius vector. The other part, depending on T, is equal to the product of T into the

---

[*] *Phil. Mag.* March and November, 1880.

resolved part of the current perpendicular to the axis of T; and its direction is perpendicular to T and to the current, being always in the direction in which the resolved part of the current would lie if turned 90° in the positive direction round T.

"Considering the current and T as vectors, the part of the electromotive force due to T is the vector part of the product $T \times$ current.

"The coefficient T may be called the rotatory coefficient. We have reason to believe that it does not exist in any known substance. It should be found, if anywhere, in magnets which have a polarization in one direction, probably due to a rotational phenomenon in the substance."

Does not the "rotatory coefficient" of resistance completely express the important facts discovered by Mr Hall? Instead of expressing these facts by saying that there is a direct action of a magnetic field on a steady current as distinguished from the body conducting the current, may we not with equal convenience express them by saying that the effect of a magnetic field on a conductor is to change its coefficients of resistance in such wise that the electromotive force is no longer a *self-conjugate*-linear-vector function of the current?

51.

NOTES ON THE SEAT OF THE ELECTROMOTIVE FORCES IN A VOLTAIC CELL.

[From the *Philosophical Magazine*, October, 1885, pp. 336—342.]

THE following is an expansion of some short remarks I made when Dr Lodge's paper was read to the Society of Telegraph Engineers.

I. *The controversy between those who hold that the difference of potential between zinc and copper in contact is what is deduced by electrostatic methods, and those who hold that it is measured by the Peltier effect, is one of the relative simplicity of certain hypotheses and definitions used to represent admitted facts.*

Taking thermoelectric phenomena alone, we are not imperatively driven to the conclusion that the difference of potential between zinc and copper is the small quantity which the Peltier effect would indicate; but by assuming with Sir W. Thomson that there is an electric property which may be expressed as an electric convection of heat, or that electricity has specific heat, we may make the potential difference as great as we please without contradiction of any dynamical principle or known physical fact. Let us start with the physical facts, and introduce hypothesis as it is wanted. These are, as far as we want them :—(1) If a circuit consist of one metal only, the electromotive force around the circuit is nil however the temperature may vary in different parts; this of course neglecting the thermoelectric effects of stress and magnetism discovered by Sir W. Thomson.   (2) If the circuit consist of two metals with the junctions at different temperatures t_1, t_2, then the electromotive force round the circuit is the difference of a function of t_2 and of the same function of t_1.   According to Prof. Tait the function is $b\,(t_2 - t_1)\left(T - \dfrac{t_2 + t_1}{2}\right)$, or, as we may write it, $A + Bt_2 - C\dfrac{t_2^2}{2} - \left\{A + Bt_1 - C\dfrac{t_1^2}{2}\right\}$; the series may perhaps extend further, but, according to Tait's experiments, the first three terms are all that are needed.

Now, but for the second law of thermodynamics we should naturally assume that $A + Bt_2 - \dfrac{Ct_2^2}{2}$ was the difference of potentials at the junction of temperature t_2, and $A + Bt_1 - \dfrac{Ct_1^2}{2}$ at the junction of temperature t_1; we should further assume that what the unit of electricity did was to take energy $A + Bt_2 - \dfrac{Ct_2^2}{2}$ out of the region immediately around the hot junction, with disappearance of that amount of heat, and to take energy $A + Bt_1 - \dfrac{Ct_1^2}{2}$ into the region immediately surrounding the cold junction, with liberation of that amount of heat.   Now apply the second law of thermodynamics in the form $\Sigma \dfrac{q}{t} = 0$, and we have

$$A\left(\frac{1}{t_2} - \frac{1}{t_1}\right) - C\frac{t_2 - t_1}{2} = 0,$$

whence it follows that $A = 0$, which may be, and that $C = 0$, which is contrary to experiment.   The current then must do something else than has been supposed, and the hypotheses differ in expression at least as to what that something else is.   The fact to be expressed is simply this: when a current passes in an unequally heated metal, there is a reversible transference of heat from one part of the metal to another, whereby heat is withdrawn from or given to an element of the substance when a current passes through it between points differing in temperature, and is given to or withdrawn from that element if the current be reversed. Sir W. Thomson proved that this follows from the fact of thermoelectric inversions and the second law of thermodynamics, and verified the inference by experiment, his reasoning being quite independent of any hypothesis.

Suppose wires of metals X and Y are joined at their extremities, and the junctions are kept at temperatures t_2, t_1. The observed electromotive force around the circuit is $f(t_2) - f(t_1)$ or within limits according to Tait, $B(t_2 - t_1) - \frac{1}{2}C(t_2{}^2 - t_1{}^2)$. The work done or dissipated by the current when unit of electricity has passed is $f(t_2) - f(t_1)$, and this is obtained by abstraction of heat from certain parts of the circuit and liberation of heat at other parts by a perfectly reversible process. Let $F(t_2)$ be the amount of heat which disappears from the region surrounding the junction t_2 when unit of electricity has passed from X to Y. Let an element of the wire X have its ends at temperatures t and $t + dt$, and let the quantity of heat abstracted from this element when unit of electricity passes from t to $t + dt$ be represented by $\phi(t)\,dt$, and let the same for Y be represented by $\psi(t)\,dt$. By the first law of thermodynamics we have

$$F(t_2) + \int_{t_2}^{t_1} \psi(t)\,dt - F(t_1) + \int_{t_1}^{t_2} \phi(t)\,dt = f(t_2) - f(t_1),$$

and by the second law, since the transference of heat from part to part is reversible,

$$F(t_2)/t_2 - F(t_1)/t_1 + \int_{t_1}^{t_2} \phi(t)/t \cdot dt - \int_{t_1}^{t_2} \psi(t)/t \cdot dt = 0.$$

Differentiating we have

$$\begin{cases} F'(t) + \phi(t) - \psi(t) = f'(t), \\ F'(t)/t - F(t)/t^2 + \phi(t)/t - \psi(t)/t = 0 \,; \end{cases}$$

whence

$$\begin{cases} F(t) = tf'(t) = Bt - Ct^2, \\ \phi(t) - \psi(t) = tf''(t) = Ct. \end{cases}$$

This really contains the whole of thermoelectric theory without any reference to local differences of potential, but only to electromotive force round a complete circuit. But when we come to the question of difference of potential within the substance at different parts of the circuit, we find that according as we treat it in one or the other of the following ways we may leave the difference of potential at the junctions indeterminate and free to be settled in accordance with hypotheses which may be found convenient in electrostatics, or we find it determined for us, and must make our electrostatic hypotheses accord therewith.

The first way is that of Thomson, as I understand it. Assume that there is no thermoelectric difference of potential between parts of the same metal at different temperatures, at all events till electrostatic experiments shall show that there is. It follows that we must assume that the passage of electricity between two points at different temperatures must cause a conveyance of energy to or from the region between those points by some other means than by passage from one potential to another. Such conveyance of energy may be very properly likened to the convection of heat by fluid in a tube, for although convection is in general dissipative, it is not necessarily so, *e.g.* a theoretically perfect regenerator. Suppose, then, that in metal X unit of electricity carries with it $\int \phi\,(t)\,dt$ of heat, and in metal Y, $\int \psi\,(t)\,dt$, this will account for the proved transference of heat in the two metals. When a unit of electricity passes across a junction at temperature t from X to Y, it must liberate at that junction a quantity of heat $\int \phi\,(t)\,dt - \int \psi\,(t)\,dt$; but the actual effect at this junction is that heat $F(t)$ disappears; hence the excess of potential at the junction of Y over X must be

$$F(t) + \int \phi\,(t)\,dt - \int \psi\,(t)\,dt \quad \text{or} \quad A + Bt - \tfrac{1}{2}Ct^2,$$

A being a constant introduced in integration. If, then, we assume a "specific heat of electricity," the actual difference of potential at a junction may contain a constant term of any value that electrostatic experiments indicate.

But the facts may be expressed without assuming that electricity conveys energy in any other way than by passing from a point of one potential to a point of different potential. This method must be adopted by those who maintain that the Peltier effect measures the difference of potential between two metals in contact. *Define* that if unit-electricity in passing from A to B points in a conductor homogeneous or heterogeneous does work, whether in heating the conductor, chemical changes, or otherwise, the excess of potential of A over B shall be measured by the work done by the electricity. This is no more than defining what we mean by the potential within a conductor, a thing we do not need to do in electrostatics. This definition accepted, all the rest follows. Between two points differing in temperature dt the rise of potential is $\phi(t)\,dt$ in X, $\psi(t)\,dt$ in Y; at the junction the excess of potential of Y over X is $F(t) = Bt - Ct^2$.

The second method of arranging one's ideas on this subject has the advantage that it dispenses with assuming a new property of that hypothetical something, electricity; but there is nothing confusing in the first method.

II. The thermodynamics of the voltaic circuit may be dealt with on either method of treatment; in the equations already used, instead of speaking only of the heat disappearing from any region, we have to consider the heat disappearing when the unit of electricity passes plus the energy liberated by the chemical changes which occur. Consider a thermoelectric combination in which there is chemical action at the junctions when a current passes.

If $G(t)$ be the function of the temperature which represents the energy of the chemical reaction which occurs when unit of electricity passes from X to Y across the junction, we have

$$F(t_2) + G(t_2) - F(t_1) - G(t_1) + \int_{t_2}^{t_1}\psi(t)\,dt + \int_{t_1}^{t_2}\phi(t)\,dt = f(t_2) - f(t_1),$$

$$F'(t) + G'(t) + \phi(t) - \psi(t) = f'(t),$$

$$F'(t)/t - F(t)/t^2 + \phi(t)/t - \psi(t)/t = 0;$$

whence

$$\left.\begin{aligned}F(t) &= tf'(t) - tG'(t), \\ \phi(t) + \psi(t) &= t\{f''(t) - G''(t)\}.\end{aligned}\right\}$$

If now we proceed on the hypothesis of specific heat of electricity, we are able to make the differences of potentials at the junctions accord with the indications of electrostatic experiments. We are, then, by no means bound in a voltaic cell to suppose that there is a great difference of potential between the electrolyte and the metal because there is a reaction there, for we may suppose the energy then liberated is taken up by the change that occurs in the specific heat of electricity.

III.   Adopting the second method of expressing the facts, we may consider further the location of the difference of potential in a voltaic cell.   In the case of a Daniell's cell consisting of

$$\text{Cu} \mid \text{CuSO}_4 \mid \text{ZnSO}_4 \mid \text{Zn,}$$

at which junction is the great difference of potential?   Dr Lodge places it at the junctions of the metals and the electrolytes.   For this there is really some experimental reason, but without such reason it is not apparent why there may not be a great difference of potential between $CuSO_4$ and $ZnSO_4$.   In that case, in an electrolytic cell with zinc or copper electrodes and $ZnSO_4$ or $CuSO_4$ as electrolyte there would exist a small difference of potential between the metal and the electrolyte.   Take the latter case, an electrolytic cell of $CuSO_4$, and let us leave out of account the irreversible phenomena of electrical resistance and diffusion. First, let us assume, as is not the fact, that the only change in the state of the electrolytic cell when a current has passed is addition of copper to one plate, loss of copper from the other plate; what could be inferred?   Imagine a region enclosing the anode; when a current has passed, what changes have occurred within the region?   An equivalent of copper has disappeared from the anode, and that same quantity of copper has departed and gone outside the region.   But by our supposition, nothing else has happened barring increase of volume for liquid by diminished volume of metallic copper; there is no more and no less $CuSO_4$ in the region, the same quantity therefore of SO_4.   All the work done in the region is to tear off a little copper from the surface of the anode and to remove it elsewhere.   If the fact were as assumed it would follow that the passage of the current did little work in the passage from copper to sulphate of copper, and consequently that the difference of potential between the two is small.   But the fact is, other things happen in the cell than increase of the kathode and diminution of the anode.   In contact with the anode there is an increase of $CuSO_4$, in contact with the kathode $CuSO_4$ disappears: this is a familiar observation to everyone.   Reconsider the region round the anode.   Assume as another extreme hypothesis that after a current has passed we have in this region the same quantity as before of copper, but more $CuSO_4$; SO_4 has entered the region and has combined with the copper.   A large amount of energy is therefore brought into the region, which can only be accounted for by supposing that the electricity has passed from a lower potential in the copper to a higher potential in the electrolyte. The legitimate conclusion is, then, that there is between Cu and $CuSO_4$ a difference of potential corresponding to the energy of combination; and the basis of the conclusion is the simple observation that the copper is dissolved off one plate but remains in its neighbourhood, whilst it is precipitated on the other plate, impoverishing the solution. In other words, it is the SO_4 that travels, not the Cu.

Now consider the ordinary Daniell's cell. Is there a substantial difference of potential at the junction of $CuSO_4$ and $ZnSO_4$? Is there, in fact, a difference apart from the Peltier difference? Imagine a region enclosing the junction in question; it might have been that the effect of a current passing was to increase the zinc and diminish the copper by an equivalent of the electricity which passed, from which we should have inferred that the seat of the electromotive force in a Daniell's cell was at the junction of the two solutions. But it is more nearly the fact that no change whatever occurs in the region in question when a current passes, and that all that happens is that a certain quantity of SO_4 enters the region and an equal quantity departs from it, from which it follows that there in no potential difference, other than a Peltier difference, at this junction.

Neither of the extreme suppositions we have made as to concentration or impoverishment of the solution is in fact true, but they serve to show that the position of the steps in potential depends entirely on the travelling of the ions. The fact is, that in general both ions travel in proportions dependent on the condition of the electrolytes; it is probable that the travelling of the SO_4 depends on some acidity of the solution. Given the proportion in which the ions travel and the energy of the *reversible* chemical reaction which occurs, and we can calculate the differences of potential at the junctions.

In the preceding reasoning an assumption has been made, but not stated. It has been assumed that the passage of a current in an electrolyte is accompanied by a movement of ions only, and not by a movement of molecules of the salt; that is, when unit of electricity passes through a solution of $CuSO_4$, xCu travels in one direction and $(1-x)\,SO_4$ in the opposite direction, but that $CuSO_4$ does not travel without exchanges of Cu and of SO_4 between the molecules of $CuSO_4$. In the supposed case when there is no concentration around the anode, my assumption is that Cu is dissolved off the anode, and that an equal quantity of Cu leaves the region around the anode *as Cu* by exchanges between the molecules of $CuSO_4$. But it is competent to some one else to assume that in this case SO_4 *as SO_4* enters the region by exchanges between the molecules of $CuSO_4$, and that at the same time a molecule of $CuSO_4$ leaves the region without undergoing any change. Such a one would truly say that there was no inconsistency in his assumption; and that if it be admitted, it follows that the difference of potential at the junction $CuSO_4 \mid Cu$ is that represented by the energy of the reaction. I prefer the assumption I have made, because it adds nothing to the ordinary chemical theory of electrolysis; but it is easy to imagine that facts may be discovered more easily expressed by supposing that an electric current causes a migration of molecules of the salt, as well as a migration of the components of the salt.

52.

ALTERNATE CURRENT ELECTROLYSIS. By J. Hopkinson, D.Sc., F.R.S., E. Wilson, and F. Lydall.

[From the *Proceedings of the Royal Society,* Vol. LIV.
pp. 407—417.]

Received November 2,—Read November 23, 1893.

Our attention has been called to the interesting work of
Messrs Bedell, Ballantyne, and Williamson on "Alternate Current
Condensers and Dielectric Hysteresis" in the *Physical Review* for
September—October, 1893. As experiments bearing upon an
analogous subject were carried out in the Siemens Laboratory,
King's College, London, we think it may be of interest to publish
them. Our experiments were commenced in June, 1892, and
were discontinued in the following July with the intention of
resuming them at a future time. They are therefore not ex-
haustive.

Suppose an alternating current to be passed through an
electrolyte between electrodes, and that the current passing and
the difference of potential are measured at intervals during the
phase. If the electrolytic action were perfectly reversible, we
should expect to find the potential difference to have its maximum
value when the current was zero, that is to say, when the total
quantity of electricity had also a maximum value. One object
we had in view was to ascertain if this were the case, and, if not,
to determine what amount of energy was dissipated under different
conditions.

This is readily done, inasmuch as the work done on the
voltameter or by the voltameter in any short time is the total

quantity of current passed in the time multiplied by the potential difference. Let a curve be drawn in which the ordinates are the coulombs and the abscissæ the volts at corresponding times: the area of this curve represents the work dissipated in a cycle.

It is, of course, well known that if a current is passed through an electrolyte, the potential difference speedily attains a certain maximum value and there remains. If an alternate current is passed, we should expect to find that as the number of coulombs passed in each half period increased, the potential difference would also increase, until it attained the value given with a continuous current, and that when this value was attained, the curve of potential and time would exhibit a flat top for all higher numbers of coulombs passed. We thought it possible that from the number of coulombs per unit of section required to bring the potential difference to its full value, we could obtain an idea of how thick a coating of the ions sufficed to secure that the surface of the plate had the chemical quality of the ion and not of the substance of the plate.

Platinum Plates.

Part I.—In the first instance, two cells having platinum plates for electrodes were used. We are indebted to Messrs Johnson and Matthey for the loan of these plates. They have each an area of 150 sq. cm. exposed to one another within the electrolyte, and are placed in a porcelain vessel $\frac{1}{8}$ in. apart. Pieces of varnished wood were placed at the back of each plate so as to prevent conduction between the outside surfaces through the fluid. The solution used was of water 100 parts by volume, and H_2SO_4 5 parts. Fig. 1

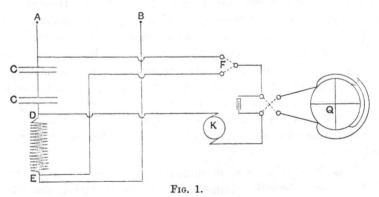

FIG. 1.

gives a diagram of connexions, in which A, B are the terminals of a Siemens W12 alternator, C, C are the cells above described, in series with which is placed a non-inductive resistance, DE. By means of a two-way switch, F, one of Lord Kelvin's quadrant electrometers, Q, could be placed across the cells C, C or the non-inductive resistance DE through a revolving contact-maker*, K, fixed to the shaft of the alternator. A condenser of about 1 m.f. capacity was placed across the terminals of the electrometer.

From observations of the values of the E.M.F. across the cells C, C at different times in a period, a Curve A (Figs. 2, 3, 4) was plotted, giving potential in terms of time.

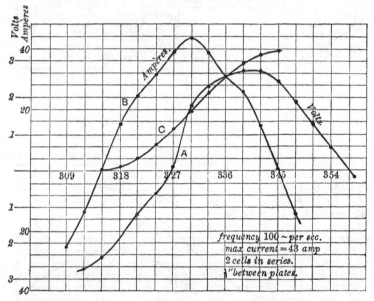

Fig. 2.

In the same way the Curve B was plotted for the E.M.F. between D and E, giving the current in terms of time. Hence the area of this Curve B up to any point, *plus* a constant, is proportional to the quantity of electricity corresponding to that point. This is shown in Curve C, which is the integral of B. The three curves, Nos. 1, 2, 3, in Fig. 5, have been plotted from Figs. 2, 3, 4 respectively, and show the cyclic variation of the

* For description of contact-maker see *Roy. Soc. Proc.* vol. LIII. p. 357.

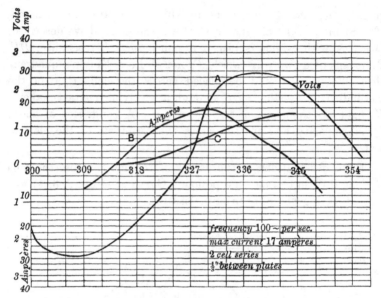

FIG. 3.

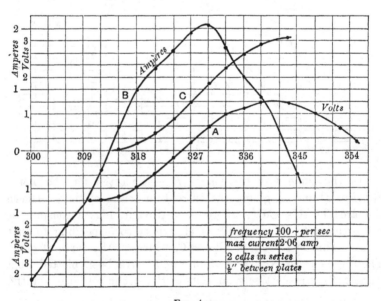

FIG. 4.

potential across the cells in volts, and the quantity of electricity in coulombs. The area of each curve (see Table I.) is a measure of the energy dissipated per cycle, and since in this case there can be no accumulation of recoverable energy at the end of the cycle, it follows that the *whole* difference between what is spent during one part of the process and what is recovered during the other part is dissipated. In order to obtain an idea of the efficiency to be looked for when used as a condenser with platinum plates $\frac{1}{8}$ in. apart and dilute sulphuric acid, under varying conditions as to maximum coulombs, the area ABC (Curve 1, Fig. 5) has been taken and is a measure of the total energy spent upon the cell; whilst the area DBC is a measure of the energy recovered—the ratio of these areas gives the efficiency.

TABLE I.

	Frequency	Maximum volts across cells	Maximum ampères	Maximum coulombs	Area of cyclic curve in diagram squares *	Efficiency per cent.
Fig. 2	100	2·7	43·3	0·065	53·8	23
,, 3	,,	2·4	17·4	0·027	9·0	24
	,,	2·38	10·0	0·0164	—	34
	,,	1·93	5·7	0·0088	—	—
	,,	1·61	2·9	0·0048	—	32
,, 4	,,	1·3	2·06	0·003	0·6	43

Part II.—In the next set of experiments the frequency was varied, in addition to current; and in order to allocate the losses of potential in the cell, the platinum plates were placed $\frac{1}{4}$ in. apart for the purpose of introducing an electrode into the fluid between the plates. This electrode consists of a platinum wire sealed into a glass tube which was capable of being placed in any desired position between the plates. The solution was, as before, of water 100 parts and H_2SO_4 5 parts by volume.

The arrangement of connexions was similar to that shown in Fig. 1, but, instead of observing the potential between the two platinum plates, observations were taken of the values of E.M.F. between one plate and the exploring electrode.

* 1 diagram square represents $\frac{1}{2}$ volt $\times 10^{-2}$ coulombs.

Table II. gives particulars of the experiments tried, and two sets of results are shown in Figs. 6 and 7, in each of which, from observations of the values of E.M.F. between the exploring electrode and the platinum plate at different times in a period, a Curve A_1 was plotted, giving potential in terms of time. This Curve A_1 is peculiar, in that the ordinates at corresponding points in the two

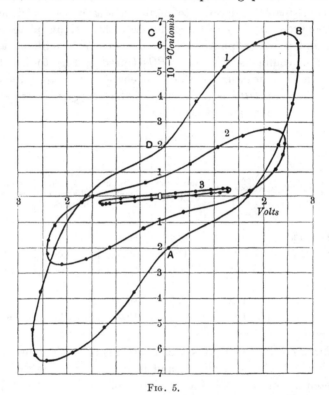

Fɪɢ. 5.

Tᴀʙʟᴇ II.

	Frequency	Maximum coulombs	Maximum ampères	Maximum volts per cell
	100	0·090	58·6	1·83
	19·7	0·082	11·2	1·57
	20·5	0·054	7·1	1·39
Fig. 6......	142·5	0·071	65·4	1·77
,, 7......	2·4	0·120	1·9	1·37

half-periods are not equal to one another, as is the case in Curve *A*, which gives the potentials across the two plates.

The Curve A_1 gives, at any epoch, the potential taken up in the evolution of gas at the surface of the plate, *plus* the potential due to the current in overcoming the resistance of the electrolyte itself. To separate these quantities experiments were made upon the resistance of the electrolyte for varying frequencies and currents. To this end the plates were placed about 2 in. apart in the fluid, and two exploring electrodes, as already described, were placed within the fluid in a straight line drawn perpendicularly between the faces of the plates, the distance between the electrodes being 4·3 cm. Some difficulty was experienced, owing

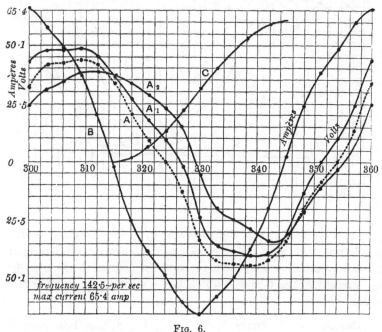

FIG. 6.

to the gases being given off at the plates more rapidly in some cases than in others. We, however, estimate that the resistance of a layer of the electrolyte, of a thickness equal to the distance between the electrode and plate, and of area equal to the area of plate submerged, in Figs. 6 and 7, was approximately 0·0056 ohm.

In Fig. 6 the Curve A_2 is the result of correcting Curve A_1 for potential lost in the resistance of the electrolyte itself, and this curve therefore gives potential taken to decompose the fluid, in terms of time. Curve B gives current in ampères in terms of time, whilst C is the integral of B and gives quantity of coulombs. With this frequency and current the energy dissipated on resistance of the electrolyte is a large proportion of the total energy dissipated; and only about 40 per cent. of the total energy is taken up in evolving oxygen and hydrogen at the plate, owing to the high frequency. The reverse of this is the case with lower frequency, as will be shown in connexion with Fig. 7.

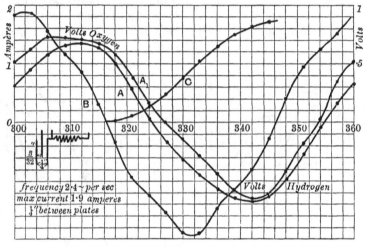

FIG. 7.

From observations on the direction in which the electrometer needle was deflected for a given position of a Clark's cell connected to its terminals, we were able to state, for a given half-period in the curves in Figs. 6 and 7, which gas was being given off at the plate.

The abscissæ of Curves Nos. 1 and 2 (Fig. 8) have been plotted from Curves A_1 and A_2 respectively in Fig. 6, the ordinates being given for corresponding epochs by the integral Curve C.

Curve No. 1 (Fig. 8) shows the cyclic variation of the potential between the electrode and the platinum plate, in terms of coulombs. Curve No. 2 shows the cyclic variation of the potential used in decomposition, also in terms of coulombs. Oxygen begins to be directed to the plate at the point A, as then the coulombs are a maximum and the current changes sign. But the oxygen is evolved on a hydrogen plate, and the E.M.F. aids the current; the work done on the plate is negative. This continues to point B (Curve No. 2). After this point (B) the character of the plate is that of a layer of oxygen and the work done becomes positive; this continues to the point C. The area AEB is the work returned

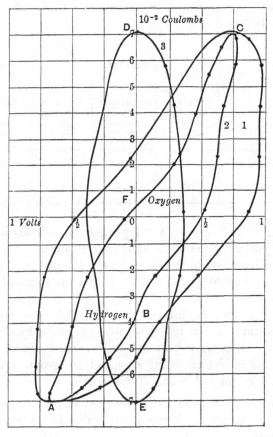

Fig. 8.

by the plate whilst oxygen is being evolved on a hydrogen surface. The area BCD is the work done on the plate whilst oxygen is being evolved on an oxygen surface. In like manner the area CDF is the work returned by the plate whilst hydrogen is being evolved on an oxygen surface, and FAE the work done on the plate whilst hydrogen is being evolved on a hydrogen surface. The above areas have been taken in square centimetres, and are given in Table III.* The area inclosed by Curve No. 2 (25·3 sq. cm.) represents the total energy dissipated by electrolytic hysteresis, whilst the area of Curve No. 1 (63·5 sq. cm.) gives the total energy spent in the cell. The abscissæ of Curve No. 3 are the differences of potential differences of Curves Nos. 1 and 2, the ordinates, as before, being coulombs. In Fig. 8, 1 sq. cm. $= \frac{1}{8}$ volt × 10^{-2} coulomb.

TABLE III.

	Oxygen on hydrogen surface, AEB	Oxygen on oxygen surface, BCD	Hydrogen on oxygen surface, FCD	Hydrogen on hydrogen surface, FAE
Fig. 8, Curve No. 2	3·65	27·25	13·8	15·5
Fig. 9	5·8	111·3	17·2	58·4

In Fig. 7 the frequency is 2·4 per second, and this is the case in which practically the whole of the energy dissipated in the cell is spent in decomposing the electrolyte at the plates. The correction to be applied to Curve A_1 for resistance is so small as to be almost negligible. The cyclic curve in Fig. 9 has been plotted from Curve A_1 and the integral Curve C, and its area (146·7 sq. cm.) represents the energy dissipated per cycle by electrolytic hysteresis. Areas have been taken in square centimetres from the curve, as in the preceding case, and are given in Table III. In Fig. 9, 1 sq. cm. $= \frac{1}{10}$ volt × 10^{-2} coulomb.

The potential curve in Fig. 7 does not exhibit a level part at the highest potential; this is possibly due to the resistance of liberated gas.

A general conclusion of the experiments is that about one-tenth of a coulomb suffices to fully polarise 150 sq. cm. of platinum. This will liberate 0·00001 of a gram of hydrogen;

* Figs. 8 and 9 having been reduced for reproduction, the absolute areas are not expressed in square centimetres, but the relative areas of the different curves are correctly expressed. [Ed.]

hence 0·00000007 gram of hydrogen serves to polarise 1 sq. cm. of platinum. 0·00000007 cm. is probably a magnitude comparable with the distance between molecules of hydrogen when this body is compressed to a density comparable with the density of liquids*.

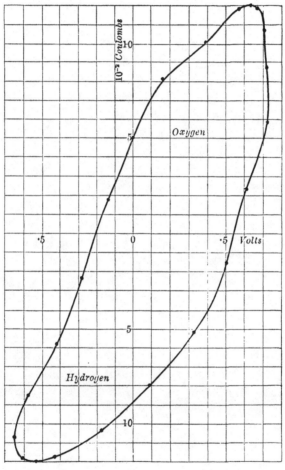

Fig. 9.

* Lord Kelvin states that in "any ordinary liquid" the mean distance between the centres of contiguous molecules is, with a "very high degree of probability," less than 0·0000002 and greater than 0·000000001 of a centimetre. See *Roy. Institution Proc.* vol. x. p. 185.